Debating Cognitive Existentialism

Value Inquiry Book Series

VOLUME 281

Central European Value Studies

Edited by

Emil Visnovský

The titles published in this series are listed at *brill.com/cevs*

Debating Cognitive Existentialism

Values and Orientations in Hermeneutic Philosophy of Science

Edited by

Dimitri Ginev

BRILL
RODOPI

LEIDEN | BOSTON

Cover illustration: Morguefile.

Library of Congress Control Number: 2015939531

ISSN 0929-8436
ISBN 978-90-04-29918-4 (paperback)
ISBN 978-90-04-29919-1 (e-book)

This book is printed on acid-free paper.

FSC www.fsc.org MIX Paper from responsible sources FSC® C109576

Printed by Printforce, the Netherlands

CONTENTS

PREFACE

Dimitri Ginev

The authors of this volume debate a version of hermeneutic philosophy called cognitive existentialism. Many years ago Georgy Markus, a brilliant philosopher from the Lukács's Budapest school, argued that there is "something contrived and artificial" in the attempts at developing a hermeneutic philosophy of the natural sciences. These attempts are simply "transposing" the readily-taken conceptions and doctrines developed in the tradition of philosophical hermeneutics to the ways in which scientific research proceeds and builds up a specific cultural field. There is no room for the celebrated figure of the hermeneutic circle in the research process even when one goes on to approach this process as being predicated on a kind of Gadamer's "logic of question and answer". (The questions arising in this process are not answered by a communicative partner, but by those who pose them, and for that reason, the questioning does not transcend the monologue of science's epistemic subject.) The figure of the hermeneutic circle stems – so Markus's argument based on Karl-Otto Apel's authority goes – from the efforts to cope with the problem of the necessity of mediation between two meaning-intentions in achieving a successful communication (i.e. one that leads to *Veständigung*) over a cognitive distance. Therefore, the hermeneutic circle is irrelevant to the monological process of scientific research. (But here comes to the fore the first question inconvenient for Markus: Was the doctrine of hermeneutic intentionalism – which so conspicuously takes place in his argument for the relevance of the interpretive circle – not expelled from the post-Diltheyan hermeneutics?)

The crucial point in Markus's line of reasoning, however, is the thesis that the natural sciences are in no need of hermeneutics. Why? For the simple reason that there is a persisting "hermeneutic naïveté" of the natural sciences' practitioners. This naïveté is not a weakness, but an indispensable moment for having a normally functioning scientific research. Natural scientists are educated (better, indoctrinated) to avoid (destructive) reflexivity that might call into question their methodological assumptions and conceptuality, and the desire to take into account the broader contexts of their work. It is the naïve attitude that brings into play the professional ideology of the natural sciences "which regards any acceptable scientific text as totally self-sufficient as to its meaning (and therefore as unambiguously clear to any reader with adequate competence)." (Markus, 1987, p. 9)

A hermeneutics of natural-scientific research would destroy this professional ideology which proves to be quite successful. Nonetheless, there is a restricted (very restricted indeed) room for practicing a kind of hermeneutics of the natural science in the form of interpretive studies of scientific communication.

More specifically, Markus places the search for hermeneutic aspects of natural science in the context of the interpretive-aesthetic theory of reception and literary hermeneutics. Accordingly, he is exclusively interested in the ways in which one confers meaning upon the *literary objectifications* of "contemporary natural-scientific discourse". Within this scope of interest he focuses predominantly on the ways of production, transmission, reception, and interpretation of the types of texts prepared for communicating scientific knowledge within various historical-cultural contexts. At stake are the "institutionalized genric-literary characteristics" of these texts that normatively circumscribe the cultural existence of science-as-literature distinguished by its canonized classical authors, tradition of reception and interpretation, audiences of readers, and spaces of authors-recipients interaction. The only function that a hermeneutics of the natural sciences thus conceived might serve is to "provide a needed corrective against often encountered biases of traditional epistemological and sociological approaches." (ibid, p. 43)

Since the publication of Markus's paper more than twenty programs of hermeneutic studies of scientific research, scientific practices, scientific knowledge, and the constitution of natural-scientific objects have been developed. A clear demarcation between interpretive constitution of meaningful entities (and orders of entities) and social construction of knowledge is drawn in these programs. The hermeneutic studies of science which implement constitutional analysis of meaning are not at odds with the conclusions which Markus drew in studying the author-text-reader relations in scientific communication. But they restore the ontological dimension of hermeneutic philosophy which Markus's approach completely ignored. Doing natural science is first and foremost a mode of being in the world. The production, transmission, reception, and interpretation of texts serving the function of communicating scientific knowledge is only a secondary aspect of this existential mode. The programs, which address science as such a mode, conceive of scientific research as facticity involved in practices and possibilities. They defy the treatment of scientific practices as an empirical presence that is directly given to theorizing in terms of (say) an empirical theory of social practices. In a hermeneutic-ontological perspective, separating practices from the possibilities they at once project and appropriate would imply the loss of the potentiality-for-being which characterizes each particular mode of being in the world. By the same token, the possibilities (for doing research and articulating scientific domains) do not have autonomous being in themselves. They only exist through their circularity with practices, i.e. in the process of their projection and appropriation. They have no being but being of changing horizons that constantly reveal and conceal aspects of what is disclosed in the research process. The circularity of practices and possibilities is still not a hermeneutic circularity in the technical sense of this expression – a mutual dependence of a meaningful whole and particular units of meaning in a process of interpretive understanding. But this circularity lays bare the grounds for specifying a diversity of constitutional hermeneutic circles that take place in scientific inquiry. Thus considered, the programs I am speaking about are (1) phenomenological in the sense of being preoccupied with the constitution of meaningful entities, and (2) hermeneutic-ontological in the sense of dealing with the circularity of practices and possibilities which grounds the potentiality-for-being of scientific inquiry.

The debate on the meaning and significance of the hermeneutic philosophy of science has mattered to many more than professional philosophers and theorists of interpretation. Over the years, this debate contributed to the ceaseless quest for figuring out a third way of philosophizing, a way beyond the depressing opposition between the analytical and Continental projects. But what is more important is that this debate has indicated strategies of transgressing the established borderlines between Snow's two cultures. By applying traditional methods of the human sciences to the constitution of natural-scientific objects within practices of research, the studies in the hermeneutic philosophy of science unveil a leeway for a dialogue between the natural sciences and the humanities, without trying to achieve a (politically) cheap "humanization of the natural sciences" that supposedly should complement the "scientification of the human studies". The leeway for a dialogue ought to enhance the radical reflexivity on both sides, and not to eliminate the multifarious tension between the (two or more) academic cultures. Hermeneutic philosophy of science is not a device for reaching *Veständigung*. It rather aims at making the disagreements productive. With regard to this aim precisely, the expression "hermeneutic philosophy of science" seems to me not quite appropriate. Studying science in the perspective of hermeneutic phenomenology, i.e. studying it as a mode of being in the world, does not try to replace (or to extend) the work accomplished in the (analytical and non-analytical) philosophy of science. In this perspective, one rather focuses on what remains unthematized in the philosophy of science, namely the existential genesis of scientific objectification (and objectivity), the pre-theoretical constitution of meaning through the manipulation of what is ready to hand in scientific practices, the disclosure of domains of inquiry, and the temporality and historicity of doing research. By addressing these subjects, "hermeneutic philosophy of science" proves to be more a critique of the philosophy of science rather than a particular version of it. "Critique" here is to be understood in a sense similar to that in which Marx develops a critique of political economy, or de Saussure treats his project of semiology as a critique of linguistics.

The debate on the hermeneutic approach to scientific research is not only a conceptual controversy. The participants in this debate bring essentially different sensibilities to their interpretation of science's cognitive, social, and cultural distinctiveness. I hope that the present volume sheds light on such sensibilities as well. At this point I should like to express my deep gratitude to Arvin Vos, Bob Crease, Fabio Gironi, Evaldas Juozelis, Jeff Kochan, and Emil Lensky for their superb critical comments and analyses of cognitive existentialism. Their contributions to this volume escalate the debates by situating cognitive existentialism in relation to a wide range of epistemological, methodological, sociological, and culturological doctrines. In my rejoinders I make use of the opportunities which the dialogue with these doctrines provides. I am also indebted to Arvin, Bob, Fabio, Evaldas, Jeff, and Emil for their accurate examinations of several weak and neuralgic points in the unfinished project of cognitive existentialism. My gratitude goes not only to the authors of this volume. More people than I can name have contributed to shaping the version of hermeneutic philosophy discussed here. I am grateful especially to two great persons, friends, and colleagues who untimely passed away — Professor John Michael Krois, and Professor Johannes Fehr. The support and good advice of Professor Emil Visnovsky, the Editor of the Series "Central European Value Studies", have been of great importance.

WORK CITED

Georgy Markus. (1987), "Why Is There No Hermeneutics of the Natural Sciences? Some Preliminary Theses," *Science in Context*, 1

Introduction

THE ROAD TO COGNITIVE EXISTENTIALISM

Emil Lensky

1. Synopsis

In 1988 Dimitri Ginev published a short paper entitled "Scientific Progress and the Hermeneutic Circle". His attention was concentrated on various aporia involved in the post-empiricist programs for a rational reconstruction of scientific knowledge's historical dynamics – aporia that eventually called into question the very idea of rational reconstruction. The starting point of the long road to cognitive existentialism was not the attempt at resolving in a hermeneutic fashion the problems which led to the normative-descriptive paradox and the vicious circularity between the epistemology of scientific progress and the (allegedly) empirical-descriptive historiography of science. Ginev's intention was rather to accomplish by means of philosophical hermeneutics an undoing of that problematic. Notoriously, the aforementioned paradox and vicious circularity brought to naught major programs of post-empiricism. These aporia are still a confusing legacy left by the historicist turn in epistemology.

The vicious circularity arises out of the search for a consistent version of science's rational reconstruction in the context of the historiography of science. This circularity consists in the impossibility to avoid the self-authentication of the (normative) philosophy of science in the history of science, granted that the latter is inevitably a normatively written history (i.e. a historiography shaped by a conception of what scientific rationality ought to be). Even the most elementary (and without any explanatory pretensions) historiographical narrative about the development (growth, change, and progress) of scientific knowledge presupposes tacitly a normative demarcation between science and non-science. Thus, all levels of historical analysis are always already enclosed in normative frameworks. Given the basic post-empiricist assumptions, a descriptive level of pure historiographical data seems, as it were, unachievable within the rational reconstruction of science. (Ginev, 1988, p. 394) But even if one were able to transform the vicious circularity into a kind of non-vicious co-dependence, thereby integrating a non-normative narrative in the scenario for a rational reconstruction, the is/ought problem and a sort of naturalistic fallacy will remain. One can neither vindicate the normative models of rational scientific behavior by appealing to purely descriptive historiographical narratives, nor "derive" a model of rationality from the narrative organization of historical data.

It is my contention that the question to which the formation of cognitive existentialism was the answer sounds as follows: Is it possible to unfold a hermeneutic alternative to normative epistemology as codified by Lakatos's three postulates of rational reconstruction? On the first of them, the task of philosophy of science consists in suggesting methodological codices in terms of which science's "internal history" can be reconstructed in a manner that would allow one to devise a "rational explanation of the growth of objective knowledge". The second postulate is the most problematic one, since it directly implies a vicious circularity. According to this postulate, "two competing methodologies can be evaluated with the help of (normatively interpreted) history." (Lakatos, 1978, p. 102) The third postulate states that the rational reconstructions of science's internal history have to be supplemented by various types of "external histories". It is this postulate that provoked several (in)famous debates between post-empiricist rationalists and those who championed the new (by the early 1970s) cognitive sociology of science.

In countering the first postulate, Ginev tried to delineate a new context of philosophical study of science that is not to be recast in terms of a kind of hybridization between the context of discovery and the context of justification. Since one handles here primarily issues of how meaningful objects become constituted in scientific research, he calls it the "context of constitution". Its delineation entangles an emphasis upon the primacy of scientific practices with the ontological turn in the Continental philosophy of science. (Babich 2007) The interrelatedness of practices in scientific research creates not only empirically approachable milieus in which all "actants" circulate. Being in a state of constant movement and being characterized by constantly changing configurations of particular practices, this interrelatedness manages to fore-structure the constitution of all meaningful entities circulating in scientific research. From the perspective of philosophical hermeneutics, the "fore-structure of understanding" does not take place in the interpreter's individual mind. It rather transcends subjective mentality, directing one's attention to trans-subjective prejudices and effective-historical traditions. Thus, the reflection upon the fore-structure of understanding amounts to reflecting on the effective-historical being (*wirkungsgeschichtliches Sein*) of the interpretative constitution of meaning. This reflection is an avenue to constructing a hermeneutic ontology. By implication, Ginev's context of constitution circumscribes a hermeneutic-ontological perspective on scientific research. The being of scientific inquiry is laid bare in this context as a being of practices projected upon possibilities.

The constitution of meaning and meaningful objects is irreducible both to cognitive-psychological processes of discovery and to processes of social construction of knowledge. But it has also nothing to do with the normative justification of the ready-made cognitive outcome of scientific research. Putting practices first (and addressing their interrelatedness in terms of hermeneutic phenomenology) helps one to avoid the vicious circularity between normative reconstruction of history and historiographical narrative. The interpretative fore-structuring of the constitution of meaningful entities lies beyond the very opposition between normative and descriptive. Yet studying scientific research within the context of constitution requires entering a hermeneutic circularity whose phenomenological analysis provides the core of cognitive existentialism. Not only the entities and actors (or, the totality of "actants") but also the methodological codices (and all other kinds of norms) are predicated on "existential genesis" when treated in terms of the context of constitution. In countering Lakatos's second postulate of rational reconstruction, cognitive existentialism draws the attention to the central "missing link" in the post-empiricist philosophy of science – the issue of the constitution of (methodological and epistemological) normativity within interrelated practices.

The fore-structuring of the interpretive constitution of meaningful entities is intimately bound up with the temporality and historicity of scientific research. More specifically, hermeneutic fore-structuring of scientific knowledge production has much to do with the regimes of temporalizing/historicizing of scientific research. "Fore" in this fore-structuring refers to what is awaiting to be appropriated as possibilities for doing research. "Fore" is a projected future that through its ongoing appropriation activates past trajectories of actualized possibilities in the current configuration of practices. As a horizon of temporality the hermeneutic fore-structuring is a futural projection that in its way of "making present" enacts what has been. Hermeneutic fore-structuring of the articulation of a domain of scientific inquiry is from the viewpoint of the phenomenology of existential temporality a kind of *gewesende-gegenwärtigende Zukunft* – "a future which makes present in the process of having been" (Heidegger, 1962, p. 401). It is this thesis that suggests a strong alternative to the distinction between external and internal history as sketched out in Lakatos's third postulate. Both kinds of history are objectivist constructions that do not take into consideration the hermeneutic roots of temporality/historicity. The latter are to be found in the ways in which the interrelatedness of scientific practices at once projects and appropriates possibilities. (Norwood Russell Hanson, Thomas Kuhn, Stephen Toulmin, Paul Feyerabend, Mary Hesse and many others shifted attention from matters of logical validation and justification to the manner in which scientific knowledge changes over time. Yet they took [historical] time as something pre-given to the epistemological studies of science's historical dynamics. By implication, the historicizing of science [not as knowledge but as inquiry] is not a proper subject of investigation in a post-empiricist perspective. For post-empiricists, historical time is a kind of "receptacle" in which different sorts of change take place. [For instance, the cumulative solving of problems by following an accepted set of norms and standards of rationality, and the Kuhnian revolutionary replacement of one paradigm by another exemplify two sorts of change somehow "embedded" in historical time.] The radicalization of historism by posing the problematic of temporalizing/historicizing of scientific inquiry is what made hermeneutic philosophy of science a unique tradition in the first place. A pioneer of this type of hermeneutic critique of post-empiricist philosophy of science is definitely Theodore Kisiel [1971, 1979 and 1997].)

In a slightly different perspective, the foregoing considerations are to be expressed also as follows. It is the projection that opens up a futural horizon for articulating a domain of research. By appropriating this horizon, one realizes the significance (*Bedeutsamkeit*) of actualized and non-actualized possibilities that have-been, thereby revealing a dynamic past that constantly depends on the projective future. What becomes present within the ongoing fusion of dynamic past and projective future does not acquire a status of static presence. It is only present as "traces of traces", i.e. in its being-scattered over a multitude of configured practices and spaces of interpretation. This sketchy outline alludes to the hermeneutic-phenomenological treatment of existential temporality in which the possibility of authentic historicity is rooted. The interpretative fore-structuring of meaningful objects in scientific research is not something that occurs "within the time". The projection and appropriation of possibilities temporalizes the interrelated scientific practices whereby the inexhaustible horizon of possibilities for doing research takes the form of a horizon or temporality. By the same token, science's historicity has no "source" but the temporality of the constitution of meaningful entities within the temporalized interrelatedness of scientific practices.

In a nutshell, what Ginev has suggested over the years is not a reformulation of the circularity in the relationship between normative epistemology/methodology and narrative historiography in terms of mutually reinforcing co-interpretation. His task is rather to show that the revealing of an ontologically relevant hermeneutic circularity in scientific research's constitution of meaning and meaningful entities would eliminate the circularity implied by the post-empiricist idea of science's rational reconstruction in a historical context. (Instead of marriage between normative theory of rationality and history of science – a marriage *prima facie* happy, but in fact full of insoluble problems – cognitive existentialism suggests a cooperation between a kind of practice theory with a hermeneutic-phenomenological analysis of meaning constitution. Ontological approaches have gained currency also in SSK, STS, feminist critique of science, and actor-network theory. In this regard, cognitive existentialism is a special case of the ontological turn in science studies.)

What is more important than the critique of the post-empiricist idea of rational reconstruction, however, is the turn to the interpretive potentiality-for-being of scientific research. A fore-structure is a contextual horizon that reveals/conceals what might be meaningfully constituted in a configuration of practices. (The whole of meaningful entities constituted in a contextualizing configuration is a relatively autonomous "text" that remains always open to a further re-contextualization.) Since the paramount outcome of research is scientific knowledge, one is to admit that all scientific practices have the status of epistemic practices. Nonetheless, the fore-structures are by no means pre-cognitive (pre-predicative) mental dispositions and attitudes that take part in the processes of shaping scientific knowledge's structure and dynamics. The ontological turn makes even scientific knowledge a "meaningful entity" within the world. Thus, scientific knowledge is incorporated in the contextualized "texts" produced by the "ontological play" of practices and possibilities. Like all other meaningful objects and artifacts, scientific knowledge is to be addressed by the constitutional analysis of meaning launched by hermeneutic phenomenology. Here is one of the special (perhaps the most important) connotations Ginev assigns to the ontological difference as employed in the context of constitution: The hermeneutic circularity fore-structures ontologically all ontic entities within the world of scientific research. (Cognitive existentialism's hermeneutic ontology of practices and possibilities bears in one important respect resemblance to the ontological assumptions of actor-network theory. In supporting the view that the trans-subjective constitution of meaningful entities [incorporated in objects of inquiry, cognitive structures, technical design, researchers, settings, etc.] dissolves any [allegedly] philosophically significant dividing line between human and non-human "actants" in scientific research, cognitive existentialism seems to subscribe to the revised [ontological] variety of the principle of symmetry as it is held, in particular, by Michel Callon and Bruno Latour [1992]. On the actor-network symmetry, whatever term one uses for humans, one should use it for non-humans as well. It is not the mutual reinforcement of two reigns clearly divided up in advance – the artificial [caused by human activity] and the natural [brought into being not by such activity] – that provides the ontological grounds for studying the networks which take place in scientific research. Rather, the activity of scientists and of all their human and nonhuman allies causes the plurality of states of nature and societies. Against the background of this view, nonhumans possess the same ontological status as the humans. Depending on the ways in which the interactions take shape, nonhumans become specified as actants that are active or passive, wild or domesticated, open or closed. Yet they never stop to be actants on a par with human actors in the research process. Of course, the exponents of actor-network approach do not deny the differences between humans and nonhumans. They treat these differences, however, as ontologically secondary. The same "derivability" of all allegedly philosophically significant human-nonhuman differences is championed in the perspective of the ontology of practices and possibilities.)

Following the lines of reasoning set up by Ginev's cognitive-existentialist strategy of overcoming post-empiricist philosophy of science, I will be preoccupied in the rest of this Introduction with three subjects (a) the hermeneutic circularity in scientific research; (b) the hermeneutic pre-normativity involved in this circularity; and (c) the pre-narrativity that arises out of the immanent historicity of scientific research. In addressing these subjects, my efforts will not be confined to the cognitive-existentialist criticisms of post-empiricism. The claim will begin to emerge that in dealing with the pre-normative grounds of epistemological norms, and the temporality of scientific practices as the "source" of science's historicity, one not only avoids both the vicious circularity and the normative-descriptive paradox but also extends the scope of philosophical hermeneutics beyond the Gadamerian borders. The reception of philosophical hermeneutics in the context of constitution is also connected with the attempt at revising established doctrines about the universality of interpretation.

2. Re-reading the post-empiricist impasse by shifting the focus from epistemology to ontology

In the heyday of post-empiricism the relationship between history of science and philosophy of science was by no means a mere marriage of convenience. In those days the conviction held sway that both disciplines are so intimately connected as to be one field. The departments of History and Philosophy of Science gained momentum. Versions of post-empiricist historism (not to be confused with the conception of "historicism" criticized by Popper) were flourishing. A prominent champion of this historism was Larry Laudan. Let me dwell on some of his arguments, from 1977, for historicizing the philosophy-of-science enterprise. According to him, creating historiographical narratives about science is always laden with implicit philosophical assumptions about the nature of science's historicity. This claim is an extrapolation of the theory-ladenness thesis and is designed to cover the ladenness of the historiographical data. Like science, historiography of science has no neutral data. Without this admission, there would be no consequent (and comprehensive) version of post-empiricist philosophy. Laudan was very cautious in working out the differences between his approach and that of Lakatos. For him, these differences are to be derived from the basic opposition between a particularist (his own) and a global version of post-empiricist historism.

Lakatos sought to formulate a model of rational reconstruction that should represent the history of science *en bloc* as a rational enterprise. Only the episodes in which the norms of rationality have been transgressed are not to be covered by such a model. For Laudan, this is a counter-intuitive approach, for it presupposes that the best model should be the one which accepts as rational every decision in the history of science that is guided by the "codex of rationality" and undisturbed by demands that contradict this codex. Furthermore, he distances his post-empiricist historism from Lakatos's quasi-Hegelian attitude expressed by the claim that by means of a rational reconstruction one tells history as it ought to have happened. Laudan's turn to approaches developed in the history of ideas is a gesture against the "rational reconstructionist" who refuses to assess the rationality of actual historical episodes. It is the history of ideas that supposedly should allow one to attend scrupulously to "the actual beliefs of historical agents". So he concludes that if the normative epistemologist "would learn something from history, he must make himself a servant to it – at least to the extent of dealing with actual cases and actual beliefs" (Lakatos, 1977, p. 170). (It deserves mentioning that Laudan fought in the mid-1970s on two fronts. He confronted Lakatos's Hegelianism, but at the same time was involved in a severe battle with the Strong Program of the Edinburgh school. At stake in this battle was the correct way of transforming meta-scientific (epistemological and cognitive-sociological) study of science into an empirical undertaking. Despite the fact that the basic positions in Laudan-Bloor controversy were subsequently overtaken, the attitudes towards the issue of how to study science scientifically are still vital and intact.

In his earlier work Ginev examined carefully this controversy. (Ginev, 1983) He came to the conclusion that a difference between ontic objectification and ontological analytic is to be made instrumental not only in the theoretical conceptuality of the meta-scientific study of scientific inquiry, but it is to be integrated in (the construction of) the appropriate empirical basis as well. It was this realization that paved the road to the context of constitution. More specifically, this context became envisaged by realizing that neither the Strong Program nor Laudan's moderate normative-epistemological historism is able to address scientific inquiry in its potentiality-for-empirical-being, i.e. to address it in a manner that excludes the assumptions of scientific inquiry's static empirical presence. Bloor and Laudan were united in the belief that "the empirically given" in the meta-scientific study of science is the outcome of methodologically constructed empirical basis. Yet they shared also the view that what is empirically given is to be identified by means of objectivist criteria pertinent to the model of deductive-nomological explanation. What Bloor counts as empirical in cognitive sociology of science is "typical sociological variables concerned with autonomy, control, prestige, status, tradition and habit". Empirically given are the transmission of knowledge, the stability of beliefs in the existence of theoretical entities, the process which go into the production of scientific knowledge, and the disciplinary differentiation of science. The requirement that the same type of explanation (or "symmetrical styles of explanation") is/are to be applied to both true and false beliefs should supposedly homogenizes the "empirical basis" of the sociological program which has to achieve the scientific self-understanding of science. Generally, the four tenets of cognitive sociology (causal explanations of beliefs, impartiality with respect to truth and falsity, symmetry with regard to the explanatory scenarios, and reflexive applicability of the same methodology to the cognitive-sociological knowledge) determine the character of the relevant empirical data. According to Laudan, the most significant flaw of the Strong Program consists in its initial postulating of a demarcation between scientific and non-scientific. He summarizes his critique in following rhetorical questions: Is it not necessary this demarcation to be studied empirically? If one is not able to depict the cognitive profile of science, how can we go on to emulate that profile? Ginev agrees with Laudan that there is a sort of *petitio principii* in the way in which Bloor's program defines its empirical basis. But he joins Bloor's party in rejecting any attempts at imposing a deductively devised theory of scientific rationality upon the methodology of studying science empirically.)

Laudan was aware that "an integrated view of history and philosophy of science" may easily encounter difficulties, the most unpleasant of which is the already mentioned vicious circularity: "If the writing of history of science presupposes a philosophy of science and if a philosophy of science is then to be authenticated by its capacity to lay bare the rationality held to be implicit in the history of science, how can we avoid automatic self-authentication, since the history we write will presuppose the very philosophy which the written history will allegedly test?" (Laudan, 1977, p. 157) In confronting this problem, Laudan firmly insisted that normative philosophy of science can dispense with historiography of science (i.e. with writing about the history of science) but it cannot escape from its dependence upon the history of science itself. With this conviction to guide him, he suggested a procedure for delineating a set of normative judgments that express the “pre-analytic normative intuitions about scientific rationality” within the beliefs about science’s progressive historical developments. These intuitions are held by “most scientifically educated persons”, and they are referring most typically to the cases of theory-acceptance and theory-rejection. He stipulated additionally that the pre-analytic judgments/intuitions thus delineated constitute a very small subset of “all our beliefs” about the pre-written (pre-interpreted) history of science. (Yet this stipulation in itself makes for a serious ambiguity: Do the pre-analytic intuitions belong to the history of science itself, or are they part – if they form a subset of “our beliefs” – of a kind of a popular historiography of science?)

Presumably, the intuitions in question might serve as touchstones for appraising and evaluating the philosophical models of normative rationality. On this account, the more of the pre-analytic intuitions a model can reconstruct, “the more confident [one] will be that it is a sound explication of what we mean by ‘rationality’” (Laudan, 1977, p. 161). But in Laudan’s eyes, the appeal to these intuitions should show a possible way out of the normative/descriptive paradox. Philosophy of science qua normative theory of rationality retains its critical function, since it provides evaluative reconstructions of the episodes for which there are no pre-analytic intuitions. He puts forward a version of Solomon’s solution, whereby philosophy of science is both descriptive and normative, characterized at once by an aprioristic-normative and an empirical status with respect to the different types of historical episodes it deals with.

Ginev criticized this suggestion for the way in which it takes for granted that methodological norms are invariably present in the Popperian “third world”, bereft of their own genesis (Ginev, 1997, 71–101; 1999a). Do not indeed these norms have their own genealogical history? Where do they come from? Do they arise from the meta-scientific clashes of epistemological conceptions about scientific rationality, i.e. not from science itself but from the attempts at its rational reconstruction? And if (explicit) methodological norms come into being in this way, do they not reflect a kind of normativity already *constituted* in (the routine everydayness of) scientific research? (By asking the final question, Ginev does not refer [like Worrall] to “implicit methodology” that supposedly remain an invariant normative core/codex, but to that interpretive pre-normativity that as fore-having, fore-seeing, and fore-conception defines the hermeneutic situation of constituting meaningful objects of inquiry. On this account, interpretive pre-normativity is the hermeneutic fore-structuring of methodological and epistemological normativity in scientific inquiry. I will discuss this topic on several occasion in the course of this Introduction.)

Obviously, if one skips the aforementioned questions, one has no chance to escape from the post-empiricist aporia discussed earlier. Ginev’s criticisms of post-empiricist philosophy were tied with the quest for a constitutional analysis (in the sense of phenomenological tradition) that can be applied to a large class of issues, all of them not receiving appropriate treatment either in the context of justification or in the context of discovery. (It is worth mentioning that with regard to Ginev's efforts to outline a new approach beyond the tradiotional context Joseph Kockelmans's criticism of post-empiricism from the viewpoint of hermeneutic phenomenology had a significant impact on cognitive existentialism [Kockelmans, 1993, Ch. 4, 5, Ginev, 2014b].)

In placing the problematic concerning the relationship between methodological normativity and historiographical narrativity in the context of constitution Ginev confronted not only Laudan's solution, but an influential criticism of this solution as well. John Worrall, the author of this criticism, argued that Laudan's approach fails to escape from the pitfall of relativism.[1] In outlining the framework in which he criticized this approach, Worrall introduced a distinction between real (implicit) methodology and explicit (declaratively announced in scientists' writings) methodology. Laudan's insistence on the mutual dependence between theoretically shaped factual claims and methodological normativity which provokes constant changes in the latter concerns only explicit methodology. By contrast, implicit methodology whose articulation had been a target of the French conventionalists as well as several logical empiricists consists in normative principles that invariably govern the appraisal of scientific theories. These are "fixed principles", and they enjoy a meta-methodological status with respect to the changeable claims of explicit methodology. There were hotly debated differences over the correct articulation of the implicit methodology's normative principles, but "there was no suggestion that two different sets of principles might be the right ones for different sets of scientists at different times" (Worrall, 1988, p. 269). On Worrall's conclusion, if there are changes in real (implicit) methodology, then Laudan's approach to methodological normativity collapses into relativism.

Ginev opposes this conclusion by stressing that the distinction between explicit and implicit methodology is at once dogmatic and (more importantly) artificial one since it is introduced by not taking into consideration scientists' reflexivity in the formation (and formulation) of the methodological principles of inquiry (Ginev, 1993). (The distinction itself dogmatically resonates the traditional context-distinction between discovery and justification.) One addresses in the context of constitution the formation and re-formation of methodological normativity with regard to contextually dependent reflexivity of scientific inquiry's participants. Due to this reflexivity, their "implicit methodology" becomes partially explicit in each particular historical context of inquiry. Doubtless, there are methodological and epistemological norms so deeply rooted in the routine of scientific inquiry that they resist explication (and specification) in science's changing normative profiles and styles of inquiry for long periods. In such periods there is a lack of research situations in which the methodological kernel of scientific rationality can be challenged and cast into doubt. Nevertheless, these norms do not form "an invariant core of meta-methodological normativity which is not subject to change and hence which is common to all the different methodologies" (Worrall, 1988, p. 274). Scientists' contextually dependent reflexivity calls into question eventually even the most stable, unproblematic and "fixed" norms. It is at the same time this reflexivity that envisions the situated transcendence of (the constitution of) methodological normativity: All methodological and epistemological norms of scientific inquiry are contextually situated in practices and spaces of interpretation, and transcended by leeway to new possible developments whose actualization might invalidate them.

Though in his critical paper on Worrall's meta-methodology Ginev invoked arguments elaborated in line with the paradigm of self-organization and autopoiesis – a paradigm from which he subsequently took a crucial distance since it naturalizes the context of constitution in an inappropriate fashion – he made it clear that the way of overcoming relativism is not to be achieved at the expense of hypostatizing meta-methodological normativity. (Studies conducted under the auspices of this paradigm are still subjected to the epistemological standards of objectivism (Ginev, 1993, pp. 331–334). Accordingly, they treat what they admit to be empirical as an actual presence even if this presence is in a constant state of self-organization and self-construction.[2] Without introducing ontological difference in theorizing scientific inquiry – so Ginev's later argument goes – one would not be able to come to grips with the potentiality-for-being [or, the being-embedded in horizons of possibilities] of what one counts as empirical in science studies. On cognitive existentialism, science [qua a cognitive reality] is empirically given in the multitude of its [epistemic] practices. To approach something that is empirically given amounts to identifying an actual state of affairs. This methodological maxim is to be applied to the multitude of scientific practices. Ginev argues that the actual state of affairs is always an outcome of actualizing possibilities. The empirically given is at once situated in possibilities and transcended by [horizons] of possibilities. This is why the actual presence of science as a cognitive reality is fore-structured by possibilities that may or may not be actualized. To study the empirically-given-in-its-situated-transcendence requires entering into the interpretative circle of the constitution of empirically manifested meaning. In studying an empirical reality not as actual presence but in its potentiality for being, one approaches this reality in its ongoing meaningful constitution. As actualized possibilities the empirical data remain constantly bound up with the horizon of non-actualized possibilities. What is empirically studied is at once discretely structured and continuously fore-structured by the iterative interpretative circle of the constitution of meaning. Studying discrete structuration within continual fore-structuring amounts to approaching the empirically manifested constitution of meaning by means of "double hermeneutics". It consists in a reflexive interpretation of the interpretative constitution of meaningful entities. This reflexive interpretation has to take into account the kinds of endogenous reflexivity involved in the interpretive constitution. The entities are meaningful for practitioners who appropriate and actualize possibilities. Being involved in the circularity of practices and possibilities, practitioners are able to make their world a meaningful reality.)

Getting rid of relativism should not imply a farewell to the idea that all epistemological and methodological norms are contextually generated. The overcoming of relativism ought to be attained by making use of (what Ginev calls) "arguments from situated transcendence". (In the context of constitution these arguments aim at displacing all kinds of relativism without assuming a "transcendental signified".) Regardless of how one were to approach the rationality of scientific research, one has to deal with the "historical reality" of the research processes consisting in piecemeal units of pursuing contextual goals and aims by following indispensable norms and standards. Basically, the functional integrity of scientific inquiry comes to the fore in the first place as a plurality of finite nexuses of norms and goals. The finitude of each particular nexus of norms and goals to be achieved is transcended by a potentially infinite horizon of possibilities in which this nexus gets articulated. The contextually situated generation of norms is "always already" transcended by such a horizon. Hence, the situated transcendence of methodological and epistemological normativity correlates with that potentiality-for-being of scientific inquiry which brings the basic assumption of any kind of (epistemic, axiological, linguistic, etc.) relativism to naught. On this assumption, the nexus of norms and goals is in its functional independence so strongly intrinsically codified that its enclosure makes it incommensurable with other nexuses.

After developing a conception of how the internal debates in post-empiricism generate the need of hermeneutic approaches, Ginev unfolded a preliminary version of cognitive existentialism. (Ginev, 1997, 2000) On its main postulate, there is nothing in scientific knowledge that is not constituted in practices of inquiry. (Put alternatively, there is pre-practical, purely contemplative or representational, layer in this knowledge.) At stake in this version is the practical constitution of science's cognitive body. Orchestrated and configured practices provide the fore-structuring in the formation of cognitive structures. Being formally codified, a holistic treatment of these structures might describe them in terms of semantic models interpreting their formalisms. Yet the practical constitution of knowledge (taking place in "scientific everydayness") requires a discourse that is capable to spell out the transition from configured practices to semantically interpretable cognitive structures. The hermeneutic paradigm of constitutional analysis was overburdened in being entitled to cope with the "jump" from routine practices of inquiry into science's cognitive structuration. The preliminary version of cognitive existentialism suffers from a conspicuous crudeness when dealing with the interpretive unity of practical fore-structuring and cognitive structuration. (Accordingly, this version did not quite successfully keep its promise to bridge the gap between science-as-practice and science-as-knowledge.) Ginev suggested a kind of interplay between the triad of fore-having, fore-sight, and fore-conception (as embedded in scientific everydayness), on one hand, and cognitive structures whose semantic models delineate a domain of research, on the other.

With regard to the issue of the transition from practical fore-structuring to cognitive structuration, this interplay which serves constitutional function can be studied in several ways (Ginev, 1999b). Since its being is interpretive-dynamic unity of practices and knowledge, the interplay acquires a status of pre-narrativity with respect to (narratives describing or making explicit) the articulation of a scientific domain. Following Ricoeur and others, Ginev conceives of pre-narrativity as a temporized-emplotted organization of facts, acts, and events that the process of narration finds "already there". The texture of scientific practices (dealing with instruments and material entities) and cognitive structures like models, hypotheses, concepts, etc. consists in organized wholes whose organization is due to intrinsic regimes of temporalizing that differentiate in characteristic manners between future, past, and present. Therefore, the temporality of these wholes is not a kind of "narrative time" – a temporal organization imposed upon scattered and amorphous units by means of an act of narration. It is rather a pre-narrative temporality that has much to do with the "temporalizing of temporality" in the sense of existential analytic.

Ginev reserves the term of "ontological narratives" for the outcome of the phenomenological analysis of the interplay (of hermeneutic fore-structuring and cognitive structures) as pre-narrativity. In contrast to historiographical narratives about disciplines or scientific domains and research programs, ontological narratives make explicit the temporality of practices that constitute meaningfully domains of specific objects. Therefore, these narratives are not *post festum* rendering of something that is already made. Pre-narrativity qua the place of scientific inquiry's temporalizing/historicizing is constantly in the making. Consequently, ontological narratives describe only possible and incomplete trajectories of domain's articulation, or in an alternative formulation, possible regimes of temporalizing/historicizing scientific inquiry. They "narrate" the pre-narrative interplay of hermeneutic fore-structuring and cognitive structuration as ongoing constitution of meaningful entities that (from the viewpoint of cognitive structuration solely) become objects of knowledge. At each moment, an ontological narrative about interrelated scientific practices that articulate cognitively a domain of research is a snapshot of an episode of scientific inquiry that through its potentiality to create futural development (i.e. to change the present trajectory of development) revises what has been. The continuity (as opposed to discreteness) and the potential infinity of such a narrative correspond to the same features of the constitution of meaning as revealed in phenomenological terms. As the outcome of a phenomenological analysis ontological narratives are interpretations of practical-interpretative fore-structuring of scientific knowledge's structuration. (Here "interpretation" means entering into the endogenous hermeneutic circularity of meaning constitution.) Thus, ontological narratives go hand in hand with a kind of double hermeneutics – the method of cognitive existentialism. (The concept of pre-narrativity underwent several specifications in the later versions of cognitive existentialism. It was related in the first place to the fore-structuring as projection and appropriation of possibilities. Since this fore-structuring temporalizes the constitution of meaning, there is a temporalized whole that exists before any emplotment achieved through narration. This whole is to be also described as a pre-narrative organization of configured practices that through their interrelatedness project possibilities and within the horizon of these possibilities appropriate them. On an alternative formulation, pre-narrativity is the iterative hermeneutic circle of meaning constitution in scientific research. The pre-narrative organization never becomes a presence by means of historiographical reconstructions since it is intrinsically instable and characterized by tensions of conflicting possibilities upon which existence projects its historical being. Historical past does not form an archive of established documental facts, but constantly depends on the appropriation of its own "futural possibilities" that the historian has to discover dialogically. It is this appropriation that retroactively brings into being past possibilities that otherwise would have had remained unnoticed. The historian is only able to study dialogically the historical event in its being projected upon possibilities, if she gets engaged in a radical reflexivity with regard to her position and hermeneutic situation of inquiry. This reflexivity concerns also the historian's situatedness in history (Ginev, unpublished manuscript).

The main deficiency of Ginev's preliminary version consists in the insufficient attention he pays to the inquiry's potentiality for being. Both scientific everydayness of configured practices and domain's cognitive structuration are addressed as ongoing, but nevertheless actually given, realities. Potentiality-for-being is attributed only to the interplay of practical fore-structuring and cognitive structures, but it remains unclear how ontological narratives take into account and thematize this potentiality. Another flaw is that this version confounds hermeneutics of mental activities and processes with trans-subjective hermeneutics of practices. I am not saying that the two hermeneutic programs should not work in a close cooperation. But in order to achieve this cooperation, one has to avoid confusing the one with the other type of hermeneutics. To sum up, the preliminary version did justice to practices of inquiry, but failed to do the same with regard to the possibilities these practices project. Nonetheless, this version tackled several issues (of prime importance to the hermeneutic philosophy of science) which unfortunately got somehow blurred in the subsequent development of cognitive existentialism. A case in point is the issue of the temporal modalities in the dynamics of scientific practices' changing configurations (Ginev, 2000, 44–55). In addressing phenomena like sedimenting, forgetting, recollecting, prolongating, and anticipating practices of inquiry, the preliminary version promised to develop a wholescale doctrine about the ways of inquiry's "historizing" and "historicizing" (both terms in the sense of hermeneutic phenomenology). The subsequent versions of cognitive existentialism dealt with facets of this problematic in a much more precise manner. Yet they have not succeeded in retaining the integral perspective on the problematic which the preliminary version put into effect. With respect to the handling of the issue of how the temporal regimes of articulating a domain of inquiry manage to engender historicity and historical time the preliminary version still succeeds in bringing its preponderance to bear.

Historicity of scientific inquiry is not something given *per se* that becomes expressed in the empirical history of science as it is rendered by historiographical narratives. The metaphysical presence of historicity is dogmatically presupposed by Lakatos's Hegelianism and Laudan's more resilient approach that invokes the history of ideas' contextualism. For Ginev, historicity (including that of scientific inquiry) is not "eternal essence" of human existence. It is engendered by the interplay of practices and possibilities within-the-world. (In fact, the phenomenologically interpreted interplay of practices and possibilities replaced in the later versions of cognitive existentialism what precisely was in my view of somewhat precarious character – the interplay of hermeneutic fore-structuring and cognitive structuration.) Historicity is not an essence to be extracted from empirical history, but an event that occurs in the temporalizing/temporalized horizon of practical being in the world. Conceived in this way, theorizing historicity of science has profound implications for the problematic discussed so far. By de-essentializing historicity (i.e. disentangling it from the metaphysics of presence), one is no longer obliged to worry about the kind of paradoxicalilty and the variety of naturalistic fallacy involved in the interrelations between descriptive historiography of science and normative-philosophical reconstruction of scientific rationality. Studying the "existential genesis" of scientific inquiry's historicity (in the context of constitution) addresses at the same time the engendering of methodological and epistemological norms in configurations of temporalizing/temporalized scientific practices (as readable technologies). Thus, by approaching the phenomena of historizing and historicizing, one also deals with the genesis of inquiry's norms of rationality out of the routine everydayness of orchestrated practices.

The post-empiricist philosophers of science are completely right in arguing that the normativity of scientific inquiry is to be only "caught" in a historical perspective. For them, this claim paves the way for a comprehensive historicizing of the context of justification. Recast in the context of constitution, the same claim implies that the hermeneutic situation (in the sense of hermeneutic phenomenology) of studying the genesis of historicity out of temporalized practices has much to do with the hermeneutic situation of studying the genesis of methodological and epistemological normativity out of practical-interpretive pre-normativity of (what Heidegger calls) "deliberative circumspection" within-the-world. In his later work, Ginev relates this pre-normativity to the triad of fore-having, fore-seeing, and fore-grasping as it gets specified by the particular configurations of scientific practices. He elaborates on a conception of "hermeneutic pre-normativity" advocated implicitly by

Wittgensteinian philosophers and ethnomethodologists as well. For him, "the uniting relationship between the individual and the social is always designed by a type of normativity (conventions, standards, regulations, prescriptions, instructions, codices, orders, sanctions, duties, obligations, maxims, prohibitions, and so on), whereas the ecstatic unity of subjectivity and trans-subjectivity is distinguished by hermeneutic pre-normativity" (Ginev, 2013b, p. 84). The latter has the character of constraints similar to those Ludwik Fleck kept in view in his conception of thought styles.

Hermeneutic pre-normativity is inherent in trans-subjectivity, i.e. in the hermeneutic circularity that transcends all subjectivity (including the inter-subjectivity of normal-scientific communities [Ginev, 2013a]). It comes to the fore as constraints informed by the triad of fore-sight, fore-having, and fore-conception in the constitution of meaningful entities. This pre-normativity is crucially tied with what Ginev calls "characteristic hermeneutic situations" – the balance and complementarity of de-contextualizing objectification and contextualizing interpretation taking place in a domain's research process. Let me shed some light on the concept of constraints. The place where this concept is discussed most widely and thoroughly is the field of organizational studies and decision making theory. More specifically, the concept is of prime importance when at issue is a process of decision making with multiple goals. Thus, if one has two goals (A and B), very often the problem situation one faces can be formulated as follows: One ought to achieve A under the constraint of B. The interesting case (in the heuristic theory of decision making) is when one is able to model the constraint (the alternative goal in the problem situation) in a flexible and integrated fashion, i.e. not as an absolute constraint (Goertz, 2004). In this case one can partially violate the constraint. Under these conditions, one has to agree with Herbert Simon's dictum that all decision making is a matter of compromise. The construal of the concept of constraints in the organizational theory of decision making revolves around the interaction between decision makers and their environment. Several theorists go so far as to accept the view that institutions are but the constraints which shape this interaction. Thus considered, institutions systematize (in the form of a certain ethos) the endogenous pre-normativity embedded in the network of interaction. In this network, the distinction between goals and constraints reveals itself to be an arbitrary distinction.

Roughly, a characteristic hermeneutic situation is to be depicted as a situation guided by two mutually constraining goals. In the language of decision making, the situation "endogenizes" the constraints. On this account, one is maximizing the de-contextualizing objectification in scientific research under the constraints of reflexive interpretation of the contexts in which this objectification takes place. By the same token, one is maximizing the "contextualizing dissipation" of the objects of inquiry under constraints of epistemological objectivity. What is at stake here is how the participants in the research process evaluate the alternatives in light of the two conflicting aims when they are choosing possibilities. It is this evaluation that is fore-structured by the hermeneutic pre-normativity. A kind of non-compensatory principle plays a crucial role in cognitive existentialism: Major "deformations" of the objects of inquiry resulting from the desire for their total de-contextualization cannot be compensated for by gains on the objectivity dimension. This principle operates in scientific research by means of pre-normative expectations, anticipations, and orientations. Thus, hermeneutic pre-normativity can be formally recast in terms of utility functions. The very process of decision making (as illustrated by these functions that relate constraints to multiple goals) transforms pre-normativity into (methodological and epistemological) normativity. I have to admit that this is not the scenario suggested by Ginev. But in my view it is the only scenario which can bind together the constitutional analysis of norms, the concept of characteristic hermeneutic situation, and the way in which cognitive existentialism lays stress upon the *choice* of possibilities.

The claim that scientific knowledge (including the stable cognitive structures) becomes constituted in scientific practices requires first and foremost a transition to a conception of science-as-practice. But in Ginev's hermeneutic perspective this claim does not entail a kind of naturalizing philosophy of science as an empirical theory of science's epistemic practices. In other words, it does not lead to the view that the meta-scientific approach to the nexus of epistemic practices and scientific knowledge is to be entirely accomplished as empirical studies (in particular, cultural studies). However, stressing the non-empirical dimensions of studying science in the context of constitution does not amount to asserting the impossibility of empiricizing/naturalizing hermeneutic philosophy of science. Ginev has tried in recent years to elaborate on a view he calls "hermeneutic naturalism" – cognitive existentialism is continuous (in a Quinean sense) with scientific research programs based on hermeneutic and phenomenological assumptions. (A slightly different formulation of hermeneutic naturalism will be discussed in the next section.) Yet this kind of empiricizing/naturalizing does not appeal to reducing the interpretive analysis of scientific practices in their potentiality-for-being to more or less objectivist studies – in which the empirical is tacitly taken to be actually given in the sense of metaphysically present – since it preserves the role which ontological difference plays in the constitution of empirical basis. To put it again in terms of hermeneutic phenomenology, Ginev alludes to a subject with many ramifications that can be headed "the empirical of objective factuality versus the empirical of meaningful facticity".

The empirical being of scientific inquiry is contextualized through scientific practices' endogenous reflexivity. Ginev distinguishes between reflexivity of practices and reflexivity of practitioners (researchers) (Ginev, 2013b, pp. 68–72). The former comes forward through the constant re-contextualization of the meaningful entities' constitution in configurations of readable technologies. Practices' endogenous reflexivity elicits the ongoing re-contextualization (of what is read and textualized) in the articulation of a domain of inquiry. A change of the context (by changing the configuration) is guided by practices' capability to refer to themselves reflexively. In their being-thrown-in-practices (and only in this being), practitioners become able to exercise their own reflexivity regarding their presumptions, basic concepts, primitive terms, operative imagery, theoretical images, etc. Consequently, practitioners' reflexivity supervenes upon practices' endogenous reflexivity. The empirical of scientific inquiry's potentiality-for-being (i.e. the empirical in the context of constitution) is the empirical of the ongoing reflexive re-contextualization. (The empirical-in-its-situated-transcendence is to be only "captured" in its routine everydayness. Each set of data is contextualized by practices and possibilities. This empirical can never become a "transcendental signified". Every contextual "identification" of the empirical leads to a re-contextualization. As a result of the actualization of possibilities the configurations of practices are constantly changing, thereby revealing new possibilities and concealing previously opened ones. By implication, the contextualized data – as a horizon-bounded empirical – are involved in the de- and re-contextualization as dimensions of the constitution of meaning.)

With regard to the ontic-ontological difference two modalities of the empirical are to be differentiated. On the one hand, it is the empirical of appropriated and actualized possibilities. It can be identified by making use of procedures from various disciplines engaged with "thick description". In this (ontic) modality, one assigns to the empirical (of scientific practices within their horizons of possibilities) the status of procedurally identifiable process and outcome of epistemic or social construction. If one goes on to ignore the potentiality-for-being of scientific inquiry, one would regard what is procedurally identified as an actual presence. Nothing more than this presence (specified, for instance, as present indexicality, or sociological indicators, or historiographical data) would be in need for studying science empirically.

The ontological modality of the empirical (as referring to the ongoing reflexive re-contextualization) is the empirical in its potentiality-for-being. To insist on the potentiality-for-being of what one counts to be empirical in science studies is not merely to revive the old view of fallibilism in a new form. One is not engage in putting old wine in new bottles since fallibilism concerns exclusively the status of the epistemic beliefs, and does not put into question the ontology of the actual (metaphysical) presence. Approaching the ontological modality of the empirical requires admitting that the actualized possibilities of scientific inquiry are contextually appropriated opportunities for doing research that are inscribed in a horizon of temporality. Through each particular contextual appropriation the horizon reveals and conceals the domain of inquiry in a specific manner. Yet it never becomes "exhausted" (or fully appropriated). Cognitive existentialism is a doctrine that takes issue with the absolutizing of the ontic modality. Despite their severe debates in the 1980s and 1990s, normative epistemologists and constructivist sociologists shared the admission that the empirical in studying science is but procedurally identifiable empirical presence. By contrast, cognitive existentialism tries to address the procedurally identifiable empirical of scientific inquiry as entangled with the projected horizon which at once situates and transcends the actualized possibilities as empirically given. On this account, the ontological modality of the empirical consists in the continuous appropriation and actualization of projected (by configured practices) horizon of possibilities which takes place in the ongoing re-contextualization of meaningful entities when articulating a domain of inquiry. The empirical in this modality is not reducible to isolable (static) facts. In taking up the motif of "ontological narrative", Ginev considers the empirical as contextualized actualizations of possibilities taking place in open horizons. Granted the continuity of re-contextualization, the empirical in its ontological (horizonal) modality is not to be expressed as discrete units of factuality, but as "temporalizing continuity of facticity" (Ginev 2007). In line with the older version of cognitive existentialism, this modality of the empirical – as representing interplaying practices and possibilities – can also be expressed as temporalizing/temporalized wholes ("stories" with plots in the making). Ginev's later conception of research-as-textualizing, which I am going to discuss in a coming section, tries to do justice to this "temporalizing continuity of facticity" of scientific inquiry. Because the horizon of possibilities which transcends each contextual actualization is a horizon of temporalization, the temporalizing/temporalized empirical of scientific inquiry becomes revealed in the context of constitution in its ongoing hermeneutic fore-structuring.

3. Taking issue with normatively shaped naturalism and constructive empiricism

In the aftermath of the controversy between post-empiricist philosophers and cognitive sociologists Joseph Rouse suggested a highly influential program that by appealing to scientific practices managed to find ways out of the descriptive-normative dilemma, thereby opening the avenue to a new brand of naturalism. Rouse's program was inspired by attempts at radicalizing the sociological approach to scientific knowledge. Cultural theorists who suggested such attempts argued that earlier cognitive sociology of science is still reifying scientific knowledge. It is a reification that analytically follows from the principle of symmetry. Scientific knowledge is regarded as a set of – true or false; rationally justified or suffering from a justificatory deficit – beliefs. The construction/production of these beliefs is subsequently explained in sociological terms. Once reified as a body of beliefs scientific knowledge (as determined socially) becomes submitted to a sociological explanation. The whole explanatory scenario based on the principle of symmetry presupposes Cartesian dualism and essentialism about knowledge. Burdened with these presumptions (or, "bad prejudices"), SSK is not capable to suggest a genuine alternative to post-empiricist historism. Rouse's initial orientation towards cultural studies of scientific practices was prompted by the task of constructing "an explicitly political philosophy of science" that by overcoming reificationist assumptions would provide "the resources for a critical assessment of the political dimension of scientific practices" (Rouse, 1987, p. 248). Accordingly, the studies of scientific practices are to be placed in an extensive political context that would allow one an assessment of the values and goals internal to scientific inquiry. What underlies this project is the sympathy for a Foucaultian, local-genealogical, unmasking of the forms of power generated by scientific practices. Ginev's attitude of simultaneous approval and severe criticism of Rouse's program had a serious impact on the formation of a more sophisticated version of cognitive existentialism.

Influenced by David Carr's phenomenological theory of narrative historicity, Rouse treated in his earlier work the interrelatedness of scientific practices as "narrative fields" Rouse, 1996, pp. 158–178). This concept was akin to the concept of pre-narrativity, and invited the introduction of ontological difference in cultural studies dealing with migration of practices across borders of scientific inquiry and ambiences in which it takes place. Later on Rouse dispensed with this concept and began to address scientific practices' interrelatedness in terms of causal interactions with the world as they are situated (as a causal intra-action) within the world. Thus, he committed to a kind of philosophical naturalism that pleads for a naturalizing philosophy of science via a causal (and normative) theory of scientific practices. (Certainly, Rouse takes into consideration "the practical situatedness" of researchers as interpreters. On Ginev's criticism, however, the situatedness which Rouse has in mind is but one's being-situated in a network of causal interactions. There is no hint at the correlation of this situatedness with the horizonal transcendence – the correlation which makes the whole network of interactions [or the whole intra-action] open to change. In Rouse's perspective – so Ginev's argument goes – the practical situatedness of interpreters implies closedness and discreteness. Presumably, only in this way one is able to identify repeatable patterns. This is not, however, the way in which Rouse identifies patterns. He advocates the anti-essentialist impetus of his position by stipulating that the pattern consists in the interactions themselves. The pattern does not emerge in the recognition or representation of a regularity. The naturalistic conception of scientific practices as expressive extensions of causal interactions with the world brings down successfully any assumption about the primacy of the epistemic cut. This conception provides enough resources for developing a version of philosophical naturalism based on "the priority of causal intra-action over regulist or regularist appeals to laws; of phenomena over objects as the locus of causal capacities; and of a normative conception of what is at stake in science over more standard appeals to the authority of objective necessity sustained by scientists' subjective commitments" (Rouse, 2002, p. 334). Nonetheless, here there is an ambiguity that has fueled Ginev's criticism over the years: Rouse's way of scrutinizing causal interactions [within the world as a total causal intra-action] tacitly admits an objectivist [and essentialist] stance towards the world. A causal network is a presence-at-hand that should be objectified in order to be investigated normatively-causally-naturalistically.)

In his book *How Scientific Practices Matter* Rouse makes the case that these practices should hold a central place in any conception of philosophical naturalism. In approaching them in terms of their normative accountability rather than their performative regularity, he combines this thesis with the view that scientific practices allow natural entities to acquire normative power over scientists' doings and sayings. As a consequence of this combination, a new, naturalistically shaped, concept of authority comes to the fore: Scientific practices (as involving material interventions that transform the world) commit one to the authority of whatever necessities and possibilities nature affords. Through this commitment – so Rouse's argument goes – nature (in its diversity of modalities) shows itself as authoritatively binding. (Science allows the natural world to be intelligible through the practices scientific inquiry deploys to interact with the natural phenomena.) The central issue is how the normative interrelations among practical interactions with the world constitute what can be understood as semantic relations (Rouse, 2002, p. 269). Stressing the philosophical primacy of causality (as incorporated in the interactions of scientific practices and natural phenomena) provides the basic rationale for the manner in which Rouse conceives of its project as a kind of philosophical naturalism. (To reiterate, scientific practices are characterized as causal interactions that disclose the natural world as consisting of causally intra-active phenomena.)

Scientific practices must be understandable as components of a system under investigation, not an external imposition upon the natural world. On this account, practitioners' intentional attribution and acknowledgement are part of the totality of system's causal intra-active phenomena. Consequently, there is no need of an anormative conception of causality entitled presumably to give an account of intentional content and associated properties. Moreover, the constitution of boundaries of total intra-actions of a phenomenal area of scientific research comes into being through the practices which express the normativity of natural phenomena. Since Rouse rejects (by means of good post-metaphysical arguments) metaphysical naturalism in general and representationalist conceptions in particular, central for his naturalist initiatives proves to be the question of how scientific understanding of the natural phenomena is to be construed if this understanding is not treated in terms of correct representation of causal necessity. Rouse aims at naturalizing philosophy of science by placing scientific practices rather than theoretical representations at the center. In so doing, he looks for a reconciliation of (in his terminology) Quinean metaphilosophical naturalism and the Nietzschean aversion to the postulation of mysterious or supernatural entities.

As I pointed out, cognitive existentialism is not an anti-naturalist doctrine. It leaves the door open for a program which Ginev dubs "hermeneutic naturalism" – naturalizing hermeneutic philosophy of science in terms of research programs in cognitive science (as natural science) that constitute their relevant empirical basis in doubly hermeneutic manner by taking into account the temporalizing/temporalized constitution of meaning. Yet cognitive existentialism is most definitely against (1) excluding ontological difference from philosophical study of science, and (2) naturalizing philosophy of science by treating the whole of scientific inquiry in terms of networks of causal interactions. With these assumptions in mind, Ginev confronts Rouse's program. Let me tentatively point out that where Rouse sees causal interactions, Ginev establishes various kinds of hermeneutic circularity. Entering the intrinsic circularity of a horizonal whole and particular units of meaning in the articulation of a research domain requires a substantial reconception of the relation between scientific inquiry and the world. This intrinsic circularity is on a par with the duality of the world which the preceding formulation assumes: On the one hand, scientific inquiry objectifies the world as a domain of research, and the world remains horizonally open in each particular objectification, regardless of how strong is the formal codification of the latter, on the other.

The contrast between Rouse's naturalism and cognitive existentialism hinges on an opposition between causal nexuses of discursive and non-discursive practices – through which the normativity of natural phenomena is expressed – and meaning-generating nexuses of co-interpretative practices-as-readable-technologies. Rouse claims, for instance, that to understand nature as disclosed through experimental interaction requires "explication of causal interaction from within" (Rouse, 2002, p. 272). A case in point is the trivial observation that a measurement interaction is intelligible only in relation to the measuring system. On Rouse's account, the concept of causal interaction inevitably involves conceptual/discursive normativity. It is the concept of causal interaction (as characterized by a normative dimension in the sense already discussed) that allows Rouse to insist on the inseparability of material and discursive interactions with the world. Scientific practices are present as regimes of causal interaction that involve interactive relations with the natural world. Ginev also admits that scientific practices are (1) normative performances, and they are (2) not externally imposed upon the phenomena under investigation. Yet, for him, (1) and (2) are not the central characteristics which a hermeneutic theory of scientific practices has to take into consideration. In this regard, he puts first the *trans-subjectivity* of the *interrelatedness* of practices. It is this horizonal interrelatedness that fore-structures (a) the causal interactions with the world taking place within the world, and (b) the normativity inherent in scientific practices.

The very commitment to a constitutional analysis provides a decisive argument against the reducibility to causal-interactionist naturalism. I would not say that scientific practices are a subject of cognitive existentialism only to the extent to which they take part in the constitution of meaning. Scientific practices are addressed in many empirical ways within this program. In cognitive existentialism, however, one makes sense of practices first and foremost in their interrelatedness which has nothing to do with a network of causal interactions. It is this interrelatedness that fore-structures both the formation of science's methodological and epistemological normativity, and the historicizing of scientific knowledge. The circular-dynamic unity of projection and appropriation of possibilities for doing research takes place within (and comes into being through) the interrelatedness of scientific practices. On this account, the latter is the "place" at which meaningful entities within scientific research become constituted. In the sophisticated version of cognitive existentialism, scientific practices are regarded as "readable technologies" (a notion introduced by Patrick Heelan), and the constitution of meaningful entities (and the articulation of a domain of research) is addressed as a process of "textualizing" (creation of "texts" through reading measurements, formalisms, data, models, experiments, etc.). Holding that there is nothing in scientific inquiry that is not predicated on potentiality-for-being amounts to claiming that there is an unceasing textualizing in scientific inquiry. Withdrawing a particular "text" (a meaningful whole constituted in a relatively autonomous configuration of readable technologies) from the process of textualizing allows one to have semantically interpretable knowledge. In the process of textualizing knowledge is a totally contextualized readiness-to-hand, being only part and parcel of what is meaningfully constituted. (In an allusion to Heidegger, this can be called "circumspective-deliberative knowledge" that circulates in configured practices.) It is the isolation of "texts" from textualizing that transforms contextualized knowledge into objective (semantically interpretable) knowledge. (The reflection upon the interrelatedness of scientific practices should give an account [in hermeneutic-ontological terms] of how the unity of projection and appropriation of horizons of possibilities constitutes meaning in scientific research. It ought to reveal the constitution of all meaningful entities that have a being-in-a-domain-of-research. Yet this reflection is by no means epistemologically irrelevant. Being fore-structured by interrelated practices, scientific knowledge proves to be also a meaningful entity. Thus, in the perspective of cognitive existentialism, the approach to the hermeneutic fore-structuring of the production of knowledge enables one to envisage a transition from ontology [of practices and possibilities] to methodology and epistemology. This is why cognitive existentialism suggests not only a shift of the problematic – from normative epistemology to hermeneutic ontology – when coping with the post-empiricist impasse, but also a critical response to post-empiricism on the terrain of methodology and epistemology. On my reading, the ingredient "cognitive" in the expression "cognitive existentialism" refers in first place to the impact which the existential ontology of practices and possibilities has on the treatment of science as a cognitive reality [including the defense of science's cognitive specificity]. In addressing this reality in hermeneutic-ontological terms, cognitive existentialism retains its status as a kind of [non-analytic] philosophy of science.)

The issue about the mode of being of science's theoretical objects acquires a special status in this version. Treating these objects as being "inscribed" in horizons of possibilities makes the transition from practices to cognitive structuration much more resilient (Ginev, 2008). Though (from epistemological viewpoint) each of them can be interpreted semantically by potentially infinite number of data models, theoretical objects do not have fixed semantic interpretation and fixed empirical content. They get their meaning contextually, within the spaces of representation of what is "textualized" through a configuration of readable technologies. Science's theoretical objects have their being through the contextual actualizations of the possibilities in which they are inscribed. They are at once in configurations of practices (and possibilities they project), and domain's cognitive structuration. Thus, by mediating between practices and knowledge, and having no being but potentiality-for-being, science's theoretical objects "participate", as it were, in cognitive existentialism's phenomenological paradigm of constitutional analysis. This move entailed a significant improvement of the preliminary version of cognitive existentialism.

One can state that always when a theoretical object gets contextually interpreted in a configuration of readable technologies, a phenomenon is saved. Like the original version of constructive empiricism, cognitive existentialism holds that "saving the phenomena" becomes possible in and through a hermeneutic circularity. Van Fraassen attributes an interpretive openness to scientific theory by admitting that the theory's models are representing (an open set of) the possible worlds allowed by the theory. In a next step, he goes on to assert that the phenomena are saved when they are exhibited as fragments of a larger unity (van Fraassen 1980, p. 56). The larger unity is the theory, but not as a complete syntactic structure. The theory can rather serve the function of saving phenomena because it allows "possible worlds" and possible semantic interpretations of its formalism. It is the intrinsic relation of mutual interpretation between theoretical terms and data that dismisses any attempt to draw the conceptual line between phenomena and the trans-phenomenal by means of a distinction of vocabulary. For van Fraassen, the hermeneutic circle of saving phenomena encloses scientific inquiry in the process of theory construction. All scientific practices are a continuation of this process "by other means". Cognitive existentialism changes the perspective: It is not the case that scientific inquiry takes place in the process of theory construction, but rather this process consists of practices that are part and parcel of the whole texture of interrelated practices of inquiry. Accordingly, the hermeneutic circle has to be formulated in a new way. The role of the "larger unity" is played by the horizonal interrelatedness of practices. Yet there is a hermeneutic circle of saving phenomena only when the interrelatedness gets specified (through a particular configuration of practices) as a contextual horizon. In this case, the mutual interpretability is between the horizon and the readable particulars.

In drawing this analogy between constructive empiricism and cognitive existentialism, I am not trying to ignore or to hide the insurmountable difference between them. What van Fraassen designates as "the hermeneutic circle in the interpretation of science" is essentially a kind of semantic interpretation (the theory-observable/measurable relation). In denying that the observable is theory-dependent, he adduces this interpretation as the way through which theory construction exhibits clear limits on observability. Though there is a proper "hermeneutic moment" in his conception – the moment of mutual interpretability between theorizing and observability/measurability that brings to naught any kind of essentialist determinism – van Fraassen basically reduces hermeneutics to model-theoretical semantics. Cognitive existentialism makes significant room for a semantic approach to science's cognitive structures. Yet, on this doctrine, the semantic interpretation (of domain's theoretical structure that delineates domain's scope and affords the intrinsic structuration of its cognitive content) can only take place when the domain gets hermeneutically disclosed by the circularity of practices and possibilities of inquiry. The relation between (ontic) semantic interpretation and (ontological) hermeneutic disclosure should be again built upon the introduction of ontological difference.

4. Confronting Gadamer's philosophical hermeneutics

The quest for making intelligible the experience of truth that transcends the scope of scientific method is the basic task of Gadamer's philosophical hermeneutics. This very formulation presupposes that "the scope of scientific method" is (in contrast to the domains of dialogical experience in philosophy, art, and history itself) not interpretatively constituted, and accordingly, the investigation of its specificity has no place in philosophical hermeneutics. All domains governed by the scientific method have nothing to do with "the whole of our hermeneutic experience". Even if one dares to scrutinize scientific research in hermeneutic terms, the outcome of this initiative would be something like a hermeneutics of writing and reading papers (and other written documents) composed for the sake of communicating experimental and theoretical results of the natural sciences. No more than this. Against the background of the tenets of *Truth and Method*, a hermeneutics of the constitution of objects of inquiry and the production of knowledge is (if not a non-sequitur or an absurd idea) at any rate an unachievable task. Gadamer argues that the "way we experience one another, the way we experience historical traditions, the way we experience the natural givenness of our existence and of our world, constitute a truly hermeneutic universe, in which we are not imprisoned, as if behind insurmountable barriers, but to which we are opened" (Gadamer, 1989, p. xxiii). Now, the question arises as to whether the exclusion of the *practical experience* involved in scientific research from this hermeneutic universe is not merely a manifestation of a "bad prejudice". For the program of cognitive existentialism, this is a central question. It is the question of how to overcome bad prejudices in universalizing the "hermeneutic problem".

Since the early 1990s Ginev has specified in several directions the motif of merging the context of constitution in philosophical hermeneutics. Yet is it possible, in a Gadamerian manner, to make sense of the very notion of hermeneutic circularity within the context of constitution? For the sake of clarity, let me dwell on how this circularity gets addressed in Gadamer's philosophical hermeneutics. At issue here is the sense in which the hermeneutic circle possesses an ontological significance. Paraphrasing Gadamer's celebrated dictum, the recognition that all understanding and cognitive articulation of the world involves some fore-structure of understanding about one's situatedness within the world gives the unfolding of an ontologically relevant hermeneutic circle its real thrust (Gadamer, 1989, p. 272). This fore-structure has a hermeneutic-ontological character because it is informed by the being of human finitude. The historical being of finite human existence – so the argument goes – becomes manifest when one scrutinizes the "fore-meanings and prejudices" informed by (the authority of) tradition. For Gadamer, there is no unconditional antithesis between the situatedness in a tradition and the freedom of reason (including scientific reason, I would dare to say, thereby countering Gadamer's view of the rationality of science).

The tradition (qua general historical horizon), on the one hand, and the particular acts of reason by means of which the tradition gets affirmed, embraced, cultivated, and eventually preserved, on the other, form a hermeneutic circularity. For the author of *Truth and Method*, a genuinely universal hermeneutics can only be achieved if one discards the view that the circular structure of understanding and interpretation is to be advocated "within the framework of a formal relation between part and whole – or its subjective reflex, the intuitive anticipation of the whole and its subsequent articulation in the parts" (Gadamer, 1989, p. 293). He argues that the hermeneutic circle is not formal in nature and consists in the "anticipatory movement" of fore-having, fore-seeing, and fore-grasping the particular meaningful entities constituted interpretatively within a projected horizon of understanding. Needless to say that in characterizing the hermeneutic circle in such a manner, Gadamer refers to the interpretative understanding of the text (a notion that refers to all entities that can be interpreted on the model of the hermeneutic circle).

As already mentioned, in his later work Ginev comes up with the idea of applying the textual model of meaning-constituting interpretation to the hermeneutics of scientific research (an area forbidden for Gadamer) as well. He makes the case that reading (and representing of what is read in the spaces of readable technologies) amounts to textualizing in scientific research. There are in the research process a lot of particular items like measurable data, outcomes of calculations, calibrated instruments, parameters of experimental systems, reports of observations, control parameters of phase diagrams, data models, theoretical models, mathematical formalisms, etc. that are read under specific conditions. No item ready-to-hand in scientific inquiry can escape the need of being read: Something exists as an ingredient of scientific inquiry to the extent to which it might be read through a readable technology of conceptualization, formalization, instrumentation, experimentation, calculation, etc. Yet neither of the aforementioned items (as represented in a particular spaces) manages to constitute an autonomous (i.e. semantically self-sufficient) text. The semantic insufficiency of what is obtained when implementing a readable technology to a particular item consist in the impossibility to define the meaning of what has been read in the same space in which it is represented (i.e. the space opened up by the item's readable technology). The meaning is "deferred" to the next space of representation in the research process. Thus, the meaning obtained by reading experimental results is to be found in the space of conceptualization, the meaning of designing a computer simulation is in the space of interpreting a theoretical model, the meaning of a measurement is deferred to the space of possible calculations within a mathematical formalism, the meaning of calibrating instruments is in the space of constructing an experimental system, and so on.

The deferral of settings of reading and spaces of representation is doubtless a deconstructionist story. Yet it has a construal in terms of hermeneutic phenomenology as well. Each item "encountered" and read within the work-world of scientific research stands in some ontological relationship towards the "worldhood" – the contextual totality of this world. The item is serviceable in its initial setting of reading. But its meaning is not to be circumscribed in this setting. The item has a meaning through its reference to something that is outside the setting. Thus, experimental data, measuring instrument, diagram, etc. are involved in the research process by means of such anticipatory references. (The totality of involvements of entities ready-to-hand in the work-world is what Heidegger calls "worldhood".) Reading a particular item in scientific inquiry allows one to "fore-see" new settings of reading/representing.

The question which now is to be asked sounds thusly: How does a semantically autonomous text arise out of semantically insufficient items? (I am saying "autonomous" and not "complete", since the requirement of semantic completeness is too strong to be fulfilled in the interpretive textualizing of scientific research. In the perspective of cognitive existentialism, semantic completeness is contextually relative to hermeneutic openness and incompleteness.) Addressing this question demands keeping in mind that readable technologies are always contextualized and configured, i.e. they are ordered not in a linear succession but in relatively closed cycles, each of them corresponding to a contextualizing configuration of scientific practices. Accordingly, the deferral of textual meaning is relatively enclosed in such a configuration/context. The meaning obtained in reading/representing particular items circulates in a cycle of deferring and overlapping spaces of representation. The totality of meaning constituted by readable technologies in a configuration of scientific practices and represented in their spaces is what Ginev calls a "text" articulated in research-as-textualizing. (Thus defined, the "text" is the meaning which circulates within the deferring spaces of representation. In the process of research-as-textualizing, a "text" does not have a presence beyond the circulation of meaning.) To say that this totality of meaning is relatively enclosed in a configuration/cycle amounts to holding that it is interpretatively fore-structured by the hermeneutic circle of reading which operates in this configuration/cycle – the circular relation of mutual dependence between the horizon of possibilities for doing research projected by the configuration and the acts of reading particular items.

The way in which this hermeneutic circle operates opens constantly new possibilities, thus enabling one to transcend the configuration of practices and to set up a new context of reading/representing/textualizing. Hermeneutic fore-structuring in research-as-textualizing is ongoing anticipation of a totality of meaning (a "text") that constantly exceeds (and in this sense transcends) each of its particular (relative) enclosures in a configuration/cycle of reading and representing. Due to this fore-structuring the process of textualizing proves to be a continuous and potentially infinite process. As I will show in the next section, because of the meaning exceeding always already each constituted "text" in scientific research – a meaning scattered over the contexts of reading/representing the "text" – the hermeneutics of research-as-textualizing is to be supplemented by deconstructionist motifs.

Because at issue in research-as-textualizing is the constitution of "texts" (and not the reproduction of meaning already constituted and incarnated in meaningful entities), Ginev puts emphasis upon the priority of textualizing (and ongoing contextualization) over the particular (relatively) ready-made texts (in their allegedly "fixed" contexts). This move implies not only a relativity regarding the enclosure of "texts" in their respective configurations/cycles, but also a relativity of the demarcation between the text (the whole which becomes read in a configuration of readable technologies) and the context (the very configuration) in scientific research – an additional point prompting the introduction of deconstructionist motifs in the context of constitution.

In dealing with the issues of research-as-textualizing, Ginev tries to enhance the role of reflexivity in philosophical hermeneutics. A line of reasoning in *Truth and Method* runs as follows. The hermeneutic sensitivity in interpretation does not require indifference with respect to the content being articulated, and accordingly, it does not require a procedure of removing the interpreter's fore-structure of understanding. This sensitivity rather demands a reflexive attitude towards the interpreter's own "fore-meanings and prejudices". Only by proceeding in this way, can the text (more generally, each particular object that the interpretative human sciences might investigate) present itself in all its otherness. On this account, however, tradition is still somehow already present before the interpretation takes place. Indeed, the tradition in which the interpreter is situated is effective, and affects the course and design of interpretation. Nevertheless, for Gadamer, the tradition is "always already" there. Of course, this is not to say that the effective-historical tradition has a "metaphysical presence". I would like rather to stress that in Ginev's eyes there is a missing element in *Truth and Method*, namely a doctrine of the co-constitution of tradition through the fusion of horizons *and* the interpreter's radical reflexivity.

Gadamer takes into account the transformation and the preservation of tradition. But this does not suffice to addressing its own interpretative constitution as a part of the effective-historical reality. Gadamer cogently argues against the romanticist conceiving of tradition in terms of something "historically given, like nature" (Gadamer, 1989, p. 282). Drawing upon his Aristotelianism, Gadamer holds that since the "real force of morals" is based on tradition, tradition is first and foremost the effective-historical ground of morals' validity. On this account, tradition is not something that operates behind the force of morals (moral normativity as *Sittlichkeit*). It is rather enacted and contextually re-enacted by this force whereby each particular re-enactment serves a grounding function with regard to morals. Here at stake is again a sort of hermeneutic circularity of contextual re-enactment and the effective-historical whole of tradition. Accordingly, the agents who respect and preserve morals are situated within tradition in a manner that allows them to recreate tradition. Now, the very same view is to be also held with respect to the interpretative investigation of the human sciences. At this point, however, Gadamer is not radical enough. More specifically, he fails to demonstrate how in each particular "research dialogue" tradition becomes – within the dialogical process – re-constituted anew. Put differently, Gadamer eschews, as it were, the issue of the sense in which the handing down of tradition in interpretative research is at the same time a re-constitution of tradition. This is the critical point which Ginev has developed over many years. It involves the need for developing (1) an extended theory of the "research dialogue" in the human sciences (including a "re-methodologization" of these sciences), and (2) a theory of the "dialogical nature" of the natural sciences (Ginev, 1995).

For Ginev, what is investigated in the hermeneutic human sciences is interpreted with regard to its contextualizing-interpretative self-constitution. Not only the objects ("texts") do not exist *per se*, but the contexts of their self-constitution are not "pre-given". In the perspective of the interpretive sciences, the contexts in which a meaningful self-constitution (of cultural life-forms, agents expressing the "spirit" of a historical epoch, cognitive and artistic styles, culturally distinguished human populations, etc.) takes place are to be thematized as inherent to the cultural objects' self-constitution (Ginev, 2007, 2014a). Like the objects, the contexts do not have a presence-in-themselves. (If one admits that the contexts are ready-made and located "forever" as boxes somewhere in the tradition – thus having a being-in-themselves – then one would not be committed to the endeavor of overcoming objectivism. Instead, one would only broaden the scope of objectivism, enriching it with a kind of "contextual objectivism".) By implication, the "texts" of the interpretative investigations are to be approached in their ability to integrate in their effective-historical being the contexts in which they achieve a meaningful self-constitution. Thus considered, the contextualizing-interpretative self-constitution of the "texts" is attainable through an interpretative dialogue with them that involves the interpreter's reflexivity. The latter scrutinizes critically (and possibly calls into question) all elements of the interpreter's fore-structure of understanding. The interpretative dialogue which re-contextualizes the interpretative self-constitution of the "texts" presupposes the interpreter's radical reflexivity. The latter precisely radicalizes

Gadamer's concept of tradition. (At the bottom of this radicalization is a conception about the "fusion of horizons" in the interpretative dialogue that regards the fusion as a stage-to-stage process in connection with the forms of the interpreter's reflexivity [Ginev, 1995].) The spirit of this radicalization is caught well in the following passage: The interpreter "is able to decide how personal experience can be effective in the present research situation, and which relevant conceptual framework and methodological pattern is more appropriate to be followed, and so on. Because of this self-reflective involvement in the trans-subjectivity of dialogical research, the interpreter is able to elaborate on a hermeneutic methodology of the human-scientific research process" (Ginev 1998, p. 264).

The quest for enhancing the role of reflection in the philosophy of the human sciences had strong repercussions in cognitive existentialism. To reiterate, Ginev's principal criticism of philosophical hermeneutics concerns the way in which the formal sciences and the natural sciences are excluded from the interpretative universe. Gadamer correctly observes that the production of knowledge in these sciences is unavoidably situated in traditions, and regardless of the degree of formalizability (in the mathematical disciplines) and objectivity (in the experimental sciences) nobody can erase the traces of human finitude in this knowledge. Despite this observation, he goes on to assert that scientific research derives the law of its development not from its hermeneutic situation but from "the law of the objects" it is investigating. (At this point Gadamer's philosophical hermeneutics is not equal to its own universalist presuppositions. Moreover, it even lags behind post-empiricist historism.) Thus, in contradicting his own assumptions (and confounding science's cognitive specificity with scientism), Gadamer ascribes an objectivist identity to scientific research and knowledge. True, in the interpretative human sciences we cannot speak of an object of inquiry in the same sense as we do in the objectifying mathematical-experimental sciences – the logic of question and answer is different. But this does not imply that we cannot treat the meaningful constitution of natural-scientific objects of inquiry from a hermeneutic point of view. Moreover, only from this viewpoint does Gadamer's insistence that scientific research "penetrates more and more deeply into nature" make sense (Gadamer, 1989, p. 285). (In his book *The Context of Constitution* Ginev insists on the distinction between "externally" opened interpretative inquiry in the human sciences and the kind of inquiry [in the natural sciences] that is able to assimilate external problems in its own horizons of possibilities for doing research. [With regard to the latter kind of inquiry he develops the concept of "hermeneutic internalism".] The externally opened inquiry is very often continuous with extra-scientific debates in the public sphere, whereas the inquiry guided by hermeneutic internalism has the potential to interiorize extra-scientific [ecological, medical, climatological, etc.] problems, thus making them problems that can be resolved by appropriating possibilities projected by practices of scientific research. If there is no external pressure exerted on scientific community, the very interiorization does not destruct or distort the intrinsic projection and appropriation of possibilities in which the freedom of scientific research is rooted. In fact, the free choice of research possibilities that are projected by the same practices in which the "choosing practitioners" are absorbed is an additional specification of cognitive existentialism that defines it as doctrine about the ethos of scientific research. [Hermeneutic internalism is the opposite of the concept of "finalization in science" – the determination of scientific research by incorporating external goals in the research process, which precludes the scientist from having free choice of research possibilities.] For Ginev, the distinction between externally opened and internally assimilative types of inquiry does not aim at resuscitating the neo-Kantian contradistinction between nomothetic and idiographic approaches/sciences (Ginev, 2006, pp.18–19). The distinction he advocates is not an epistemological one at all. Furthermore, it does not coincide with any variety of the "two cultures" conception, though it might significantly contribute to initiatives for a "hermeneutic unity" of scientific culture that opposes any methodological or epistemological unification of science. On Ginev's account, the problematic of this unity ought to be addressed in in terms of the context of constitution. The struggle for a hermeneutic unity of science is not only a gesture against post-empiricism. This unity is also an alternative to the forms of unification which gained momentum with the rise of social constructivism and neo-pragmatist philosophizing.)

Hermeneutic phenomenology of the natural sciences looks like both highly esoteric and too exotic initiative. Yet this initiative proves to be indispensable for everybody who champions the post-metaphysical universalizing of hermeneutics. Without approaching the interpretative nature of the natural sciences philosophical hermeneutics would be essentially restricted. Without scrutinizing scientific research as a reading process distinguished by its own effective-historical being, one would have to refrain from laying a hermeneutic claim to the conceptually most sophisticated form of culture – that of objectifying the world by means of research practices that appropriate the same possibilities which they project as horizons of doing research. Rejecting the interpretative nature of the natural sciences would amount to asserting (in a quite anti-hermeneutic manner) that the "mode of being as scientific research" is not dependent on human finitude. Such a rejection would contradict the central argument of hermeneutic philosophy: It is the finitude of human existence that necessitates the interpretative character of all human practices and activities. Let us remember that Gadamer's philosophical hermeneutics is grounded upon a phenomenological philosophy of finitude and temporality. It is the ontology of finitude that allows Gadamer to argue for the primacy of the interpretative reasonableness of practical knowing and the universality of *phronesis* and dialogue. The "universal problem of interpretation" would not be resolved, if one would go on to disfranchise *phronesis* and practical rationality on the territory of natural-scientific research. The way of addressing this problem includes the task of overcoming gradually (in the sense of Heidegger's *Verwindung*) the normative-epistemological (and methodological) concept of reason. Gadamer's philosophical hermeneutics of finitude universalizes the dialogical interpretation by recognizing the limits of all knowledge which is bounded by its own historical situation. The constitution of situated knowledge cannot escape the fusion of horizons that constantly transcends each current situation. By the same token, the natural-scientific reading of theoretical objects which constantly textualizes in spaces of representation through implementing readable technologies cannot escape the horizonality of its own practices.

5. Deconstructionist motifs

Ginev's elaborations of the topic of texts' relative semantic autonomy run against structuralism in twofold sense. First, he opposes the traditional structuralist methodology (in the human sciences) which postulates that a text (regardless of how semiotically it is designed) has a hidden structure of codified meaning. Second, cognitive existentialism takes issue with basic conceptions of scientific structuralism qua a position in the realism debate. Adherents to the traditional methodology argue that there is a room for structural analysis (as it is employed in literary theory, ethnology, or history of art) in the philosophy of the natural sciences. On their account, a scientific community constitutes wholes (as analogs to texts) involving theoretical concepts, theoretically laden data, rules of correspondence, formal techniques, etc. whose meaning is codified in community's "collective mind". Structural analysis de-codifies the meaning of such a whole as a leeway of possible (explicit) variations of an (implicit) structure. The de-codification is in the first place a semantic interpretation of the wholes as analogs to texts. Yet structural analysis aims at a "deeper layer". Since the meaning of a whole is codified in a community's collective mind, the de-codification of the implicit structure should reveal how this structure supposedly determines community's behavior in the research process. Needless to say that this view of structural analysis advances logocentric essentialism and semantic determinism. According to Ginev, scientific structuralism suffers from the same flaws. Champions of this position are prone to claim that epistemic representation by means of structure is a worthwhile pursuit in its own right. They leave open the question as to whether the very representation is one of structure. (Depending on how this question is tackled, the epistemic and the ontic versions of scientific structuralism have become differentiated.) But despite the growing diversification of such versions, scientific structuralists are united and quite definite in asserting that the "mediating text" (the structure that mediates representation) makes the world a pure presence – a step that one needs in prioritizing the semantic interpretation of scientific theories. (Indeed, for advocating scientific structuralism is neither necessary to privilege the issue of epistemic representation nor to bind this position with a semantic view of scientific theories. Nonetheless, all versions of scientific structuralism approach the world as a pure presence.)

I mentioned in the preceding section that – according to the doctrine of research-as-textualizing – a "text" is a whole which becomes read in a configuration of scientific practices, whereas the very configuration builds the context. The whole which is read in such a configuration is tantamount to the total meaning of items (as data, experimental facts, measurements, diagrams, computer simulations, computations, etc.) circulating in the readable technologies' deferring and overlapping spaces of representation. By putting textualizing first, cognitive existentialism aims at undoing any stable demarcation line between text and context. (A tendency toward fusing text and context is to be observed in many scientific branches in which practices of inquiry as readable technologies become part of what is constituted through reading – the objects of inquiry. In this regard, cognitive existentialism is congruent with science's intrinsic developments.) Let me concretize the sense in which cognitive existentialism weakens and relativizes the demarcation in question. Several approaches to literary criticism stress that the context is always integrated in the text. Yet the doctrine of research-as-textualizing advocates a much stronger deconstructionist thesis: There is in scientific research a constant interplay of outcomes and tools of reading that makes the text-context a dynamic unity that exists as a non-presence in a configuration of readable technologies. To be sure, this interplay strongly resemblances – as Hans-Jörg Rheinberger shows in particular – the interplay of signified and signifier, which opens the door to a broader implementation of deconstructionist ideas in the context of constitution. (Here I am referring to deconstruction exclusively as a mode of reading. Following Derrida's rule that deconstruction uses the very principle it deconstructs, the cognitive existentialist states that textualizing – by means of reading/representing in scientific inquiry – deconstructs the texts it constitutes contextually.)

The first deconstructionist motif that supplements hermeneutics in cognitive existentialism is that of intertextuality. Relativizing the text-context border – whereby the text ceases to be a static body and becomes disseminated in the contextualizing configuration – and demonstrating that each particular configuration of readable technologies (and their spaces of representation) always already contains (and is contained by) other configurations allow Ginev to bring the concept of intertextuality (inter-contextuality) to bear in studying research-as-textualizing: Not only the meaning of the particular items within a configuration is disseminated in the deferring spaces of representation, but the whole meaning of the particular "texts" (in their relative semantic autonomy) is scattered as "traces of traces" over configurations of scientific practices. To put it in Ginev's words: "The meaning of a particular text is not a presence that waits for its explication… The meaning of a text is to be sought in relation to time as difference, differing, and deferral" (Ginev, 2009, p. 181). Invoking intertextuality in the context of constitution is a generalization of the idea that there is no meaning in the research process beyond the meaning circulating in scientific practices. In this regard, intertextuality is conceived as a deconstructionist envisioning of the research process' situated transcendence.

Poststructuralist critics insist that what is deconstructed in deconstructive criticism is not the text itself but the text as it is read. Ginev brings this claim in line with the view that the "text" remains constantly open due to its continuous hermeneutic fore-structuring. It is the convergence between the hermeneutic view of the priority of textualizing over allegedly completed "texts" and the motif of intertextuality that enables the advancing dialogue between philosophical hermeneutics and deconstruction in cognitive existentialism.

NOTES

1. Worrall criticizes the so-called "reticulated model" of Laudan (1984, 50-61).
2. For a similar argument see Janich (1992).

WORKS CITED

Babich, Babette. (2007) "Continental Philosophy of Science." In *The Edinburgh Companion to the Twentieth Century Philosophies*, ed. Constantin Boundas. Edinburgh: Edinburgh University Press, pp. 545–558.

Callon, Michel and Latour, Bruno. (1992) "Don't Throw the Baby Out with the Bath School!" In *Science as Practice and Culture*, ed. Andrew Pickering. Chicago and London: Chicago University Press, pp. 343–368.

Gadamer, Hans-Georg. (1989) *Truth and Method*. London and New York: Continuum.

Ginev, Dimitri. (1983) *The Post-Positivist Tradition*. Sofia: Bulgarian Academy of Sciences. (In Bulgarian).

———. (1988) "Scientific Progress and the Hermeneutic Circle," *Studies in History and Philosophy of Science*, 19:3, pp. 391–395.

———. (1993) "Do We Need Fixed Methodological Principles?" *British Journal for the Philosophy of Science*, 44:2, pp. 329–334.

———. (1995) *Die Mehrdimensionalität geisteswissenschaftlicher Erfahrung*. Essen: Die Blaue Eule.

———. (1997) *A Passage to a Hermeneutic Philosophy of Science*. Amsterdam, Atlanta: Rodopi.

———. (1998) "Rhetoric and Double Hermeneutics in the Human Sciences," *Human Studies*, 21:3, pp. 259–271.

———. (2000) *Critique of Epistemological Reason*. Sofia: Bulgarian Academic Monographs.

———. (2006) *The Context of Constitution. Beyond the Edge of Epistemological Justification*. Dordrecht: Springer.

———. (2007) "Doppelte Hermeneutik und Konstitutionsanalyse," *Deutsche Zeitschrift für Philosophie*, 55:5, pp. 679–688.

———. (2008) "Cognitive Existentialism," *Iyyun. The Jerusalem Philosophical Quarterly*, 57:2, pp. 227–242.

———. (2009) "Text and reading in scientific research," *Divinatio. Studia Culturologica*, 29:1, pp. 179–186.

———. (2013a) "Hermeneutic Pre-Normativity," *Philosophical Forum*, 44:3, pp. 275–293.

———. (2013b) *Practices and Possibilities*. Würzburg: Königshausen & Neumann.

———. (2014a) "Diltheyan Varieties of Double Hermeneutics in the Human Sciences," *Philosophy of the Social Sciences*, 44:1, pp. 112–120.

———. (2014b) "The Universality of Hermeneutics in Joseph Kockelmans's Version of Hermeneutic Phenomenology." In *The Multidimensionality of Hermeneutic Phenomenology*, eds. Babette Babich and Dimitri Ginev. Dordrecht: Springer, pp. v-xiv.

———. (Unpublished manuscript) "Heidegger's Concept of 'Authentic Historical Science'".

Goertz, Gary. (2004) "Constraints, Compromises and Decision Making," *Journal of Conflict Resolution*, 48:1, pp. 14–37.

Heidegger, Martin. (1962) *Being and Time*. Translated by John Macquarrie & Edward Robinson. San Francisco: Harper Collins Publishers.

Janich, Peter. (1992) "Die methodische Ordnung von Konstruktionen. Der radikale Konstruktivismus aus der Sicht des Erlanger Konstruktivismus." In *Kognition und Gesellschaft*, Hg. Siegfried Schmidt. Frankfurt am Main: Suhrkamp, pp. 24–41.

Kisiel, Theodore. (1971) "Zu einer Hermeneutik naturwissenschaftlicher Entdeckung," *Zeitschrift für allgemeine Wissenschaftstheorie*, 2:2, pp. 195–221.

Kisiel, Theodore. (1979) "The Rationality of Scientific Discovery." In *Rationality To-Day*, ed. Theodore F. Geraets. Ottawa: Ottawa University Press, pp. 401–411.

Kisiel, Theodore. (1997) "A Hermeneutics of the Natural Sciences? The Debate Updated," *Man and World*, 30:3, pp. 329–341.

Kockelmans, Joseph J. (1993) *Ideas for a Hermeneutic Phenomenology of the Natural Sciences*. Dordrecht: Kluwer Academic Publishers.

Lakatos, Imre. (1978) *The Methodology of Scientific Research Programmes. Philosophical Papers. Volume 1*. Edited by John Worrall and Gregory Currie. Cambridge: Cambridge University Press.

Laudan, Larry. (1977) *Progress and Its Problems: Towards a Theory of Scientific Growth*. Berkeley and London: University of California Press.

Laudan, Larry. (1984) *Science and Values: The Aims of Science and Their Role in Scientific Debate*. Berkeley and London: University of California Press.

Rouse, Joseph. (1987) *Knowledge and Power: Toward a Political Philosophy of Science*. Ithaca: Cornell University Press.

Rouse, Joseph. (1996) *Engaging Science: How to Understand Its Practices Philosophically*. Ithaca: Cornell University Press.

Rouse, Joseph. (2002) *How Scientific Practices Matter*. Chicago: University of Chicago Press.

van Fraassen, Bas. (1980) *The Scientific Image*. Oxford: Oxford University Press.

Worrall, John. (1988) "The Value of a Fixed Methodology," *The British Journal for the Philosophy of Science*, 39:2, pp. 263–275.

One

ON THE STATUS OF THEORETICAL OBJECTS IN SCIENCE ACCORDING TO COGNITIVE EXISTENTIALISM

Arvin Vos

In recent work Prof. Dimitri Ginev has been developing a view of the nature of science which he calls "Cognitive Existentialism." These subtle and detailed analyses call for much more extensive comment than I could possibly make here. Accordingly, I am going to limit myself to examining a single, fundamental issue: the status of theoretical objects in science. As Ginev states, the "doctrine about the status of science's theoretical objects constitutes the kernel of cognitive existentialism" (Ginev, 2009, p. 382). Again, he elsewhere states that his goal is to describe "the 'immanent transcendence" of science's theoretical objects" (Ginev, 2006, pp. 124–131). Both the immanence and the transcendence of theoretical objects are significant: the *immanence* because it is Ginev's means of countering essentialism and *transcendence* because it is necessary to account for basic features of scientific research.

My comment will have three parts: first, a brief account of the nature and status of theoretical objects in Ginev's account of science; second, an account of some difficulties that I find with Ginev's account of "theoretical" objects; thirdly, it will be argued that these difficulties have their source in the fact that implicit in Ginev's position is an understanding of knowing that is essentialist with the result that he eliminates significant features about knowing in order to avoid essentialism. Finally, I will argue that there is an alternative account of knowing which avoids the problems of essentialism and is more adequate in accounting for the facts of knowing, specifically the role of theoretical objects in science. This claim will be argued first by sketching briefly an alternative conception of the origin, nature, and role of theoretical objects. Secondly, it will be shown that this alternative approach has the capacity to resolve the issues that remain puzzling in Ginev's account of the constitution of science.

In short, I shall argue that Ginev has failed to break cleanly with the fundamental assumptions about knowing on which essentialism is based and the result is that his own account of the role of theoretical objects in science is insightful, but ultimately not fully satisfying.

1. On Science's Theoretical Objects

Cognitive existentialism aims to give an account of theoretical objects in science. The problem, as Ginev sees it, is to give an adequate account of the role of theoretical objects as they function in science without lapsing into some form of essentialism. Positively, for Ginev this means describing the role of theoretical objects in science purely in terms of scientific practices. Negatively, it means avoiding all implications that theoretical objects are cognitive essences: "There is no 'essence' behind or beyond the interrelatedness of practices" (Ginev, 2009, p. 366).

How is meaning constituted? Or, to ask the question in another way: how do theoretical objects arise? Ginev's answer is that it is through "appropriating (actualizing of) possibilities through interpretation." Again, it is said that scientific research, a mode of being-in-the-world, "projects its existence upon possibilities." How does this occur? It is through this appropriation that "an ongoing articulation of a domain's research objects comes into being" (Ginev, 2009, p. 381). *Appropriation* and *articulation* are two ways of describing how a scientist applies a theoretical object to a new situation, a new context, to another instance or instances of what was before only a possibility, something not yet known. An added aspect is that theoretical objects are "constantly undergoing small changes in normal scientific research due to their 'recontextualization' in new configurations of practices" (*ibid.*). In other words, the range of data to which a theoretical object is applied constantly expands. These statements document Ginev's sensitivity to the contextual nature of scientific reflection and the incremental development that is characteristic of normal science.

But what is appropriation? How does it occur? According to Ginev scientific research is "a mode of being-in-the-world" through which Dasein projects its being upon possibilities. Also, through appropriation "a domain's research objects come into being". Both aspects are significant and will need to be examined in more detail. However, in this context Ginev is quick to affirm the anti-essentialist nature of his position: "the ongoing hermeneutic fore-structuring of [a] domain's cognitive structure is not behind or beyond the interrelatedness of practices" (Ginev, 2006, pp. 135–139). The very fact that Ginev feels the need to reaffirm the anti-essentialist character of cognitive existentialism suggests that the matter is not entirely settled in his mind. In any case, the denial of essentialism does nothing to explain the origin of this cognitive structure, its dynamics, or its organizing principle.

Why such explanation is needed can be seen from the fact that Ginev adds an account of the nature of this cognitive structure. He notes that "a domain's research objects are always related to theoretical objects that are not present at hand in normal scientific everydayness" (Ginev, 2008, p. 237). The problem is to explain how research and theoretical objects are related. For Ginev a domain's "research objects" seem to be the individual instances or the materials that are found in the laboratory or are the things being studied in a research program. In other words, research objects are the particular chemicals, even the individual sample, which a chemist uses in his work, or the particular plants which the biologist studies when studying the growth of wheat for example, etc. In contrast to these specimens "theoretical objects...are not present at hand." Rightly, it is recognized that they have a different status. Ginev goes on to explain that "the theoretical objects always already transcend the actual configurations of practices." This fact leads Ginev to conclude rightly that there is a "'content' of these objects that cannot be exhibited by the models constituted by the configurations in progress" (Ginev, 2009, p. 382).

What is this "content" and how does it arise? For Ginev there is no simple answer. What seems to concern him most is to explain is that this admission of a "content" does not open the door to admitting "'cognitive essences' which are...independent of the dynamics of practices of normal scientific research [and which] even determine this dynamics". Research practices, we are told, do not serve the function of "'operationalizing' theoretical objects" (Ginev 2009a). The Platonism (and hence essentialism) of this rejected position is readily apparent.

But clearly it is not enough to say what theoretical entities are not; one must say what they are, how they arise, and how they function. And so we need to follow Ginev's account closely. Instead of being essentialist in character, "theoretical objects do not exist *per se*..." but have a "potential being." This potential being stems from the fact that they can be identified "through the ongoing actualization of possibilities from the horizon of projected possibilities" (Ginev, 2008, p. 234). Another way in which the potential character of these entities is described is in terms of having "both a 'visible existence' (actualized theoretical possibilities under certain experimental conditions) and 'invisible existence' (possibilities that might never be 'appropriated by the configurations of scientific practices)" (Ginev, 2009, p. 382). Again, science's 'theoretical objects' have potential being because they "project its [science's] existence upon possibilities" (Ginev, 2008, p. 236). For Ginev this potential-being claim about the nature and role of theoretical objects in science is the fundamental claim of cognitive existentialism.

Ginev explains the status of theoretical objects in more detail. Science's theoretical objects are "predicated on a double status"—as "embedded in the hermeneutic fore-structure of scientific research and in the structure of a scientific domain" (Ginev, 2009, p. 380). The latter claim is fairly obvious, for theoretical objects—here I would prefer "theoretical terms"--constitute the conceptual structure of every field of science. The first point is both more significant and more difficult. What is meant by the "hermeneutic fore-structure of scientific research?" Ginev states that theoretical objects are "'partially' present at hand and ready to hand in research practices." His explanation of this status is that they exist "through various anticipations, expectations, and orientations" (Ginev, 2006, p. 137). Why theoretical objects are only "partially" present at hand and ready to hand is not explained.

A further clarification concerns the means by which theoretical objects are grasped. It is asserted that a transcendental reflection is required in order to grasp the status of these theoretical objects. The situation is parallel, according to Ginev the reflection on the hermeneutic circle. "The status of science's theoretical objects is "revealed by the same transcendental reflection that unfolds the hermeneutic circle between fore-structure of interpretation and explicit structure as the very circularity is mediated by the interrelated practices of normal scientific research" (Ginev, 2009, p. 383). This last claim is the heart of Ginev's position. Through a transcendental reflection he believes it is possible to show that the practices mediate the relation between a scientist's anticipations, expectations, and orientations and the explicit structure, the theoretical objects, of his field of research.

The result, according to Ginev, is a situated transcendence. Situated transcendence is not "empirically given, but it conditions the empirical dynamics of scientific research" (Ginev, 2009, p. 383). Such reflection reveals the self-referentiality of scientific research, "that one understands oneself with respect to the possibilities...already projected in one's practices" (Ginev, 2009, p. 384). Again, if I engage in scientific research "the choice and appropriation of possibilities" are my actions, so have "reference to...[my] being-in-the-world" (Ginev, 2006, p. 51). Thus, "a structure of self-referentiality" coincides with the "structure of meaning constitution." Because the same will be true for other researchers, self-referentiality is a "trans-subjective phenomenon," because of the situated transcendence in which each finds oneself.

Finally, for Ginev this way of treating self-referentiality leads to a "double hermeneutics"—"the process of interpretative constitution of meaning on the level of the empirical dynamics of practices," and the "moment of situated transcendence in this dynamics" (Ginev, 2008, p. 235). Opposing what he sees as an insurmountable dichotomy in Heidegger's ontico-ontological difference, Ginev insists that there is no "autonomous transcendental-ontological knowledge." The only adequate account is based on "contextual transcendental knowledge" (Ginev, 2009 a).

2. About Theoretical Objects

Ginev's subtle and nuanced account invites thoughtful response. Two concerns dominate his discussion: on the one hand, there is his concern with avoiding any hint of essentialism; on the other, the contextualizing all constitution of meaning in practices is seen as the way to accomplish this goal. I find myself in agreement with Ginev's goals. I agree that essentialism has many problems and must be rejected. And I also agree that the constitution of meaning must be contextualized. While agreeing with Ginev's goals, I have reservations about his solution, for it appears to me that Ginev has not broken completely and cleanly with the basic assumption in the essentialist account of knowing. While he rightly rejects the results of an essentialist view of knowing, rejecting the metaphysics, epistemology, and conception of objectivity found in this view, he fails to identify and critique the fundamental misconception that is the source of the many problems in essentialism. As a result he remains caught in a number of issues that are found in the essentialist framework. What Ginev fails to critique are the essentialist's assumptions concerning the nature of knowing. This failure has multiple consequences, especially for his account of the status and role of theoretical objects. It appears to me that because Ginev fails to discover a more adequate account of knowing, he is forced to account for theoretical objects in science by focusing solely on practices. Or to put the point in another way, he insists that in his view the theoretical objects of scientific research have only an immanent transcendence, but he fails to show how this is not a kind of essentialism—in spite of protestations that his view is not essentialist. In short, in order to avoid essentialism—the view that in some way ideas exist apart from mind—he describes theoretical objects purely in relation to practices. In this section I will note some of the problems that it seems to me remain unsolved in Ginev's account of science.

Practices have a significant role in science, and so Ginev rightly focuses on them but nowhere does he indicate how practices give rise to theoretical objects. To say that theoretical objects arise in the context of practices is not enough. One wants to know why they arise in this context and how? The question cannot be avoided because clearly such objects are necessary to account for the generality of scientific knowledge. To put the point in another way, they are required to explain both understanding and interpretation, both that they are generated in a particular research situation and that they identify a range of possibilities that can be actualized in an indefinite number of other settings. It is awareness of these factors that lies behind Ginev's affirmation that, unlike research objects, theoretical objects are not present at hand.

A second question also relates to theoretical objects, but here the focus is on the fact that they go beyond "configurations of practices" (Ginev, 2006, pp. 75–92). When one reads that a theoretical object goes beyond configurations in progress, one immediately suspects that a theoretical object is "an invariant theoretical object," which is precisely what the essentialist claims. How is the situated transcendence of these theoretical objects different from the essentialism that Ginev is opposing? If theoretical objects are not "present at hand" and in this way are different from research objects, and they "transcend the actual configurations of practices," is this not exactly what the essentialist is also claiming? Ginev does not tell us how he avoids this conclusion. To say that they are discovered in the context of practices is not enough. Even Plato in the *Meno* asserts the same. Plato underscores the need for inquiry, but insists that what is discovered is a form or idea that is known in a prior life. It is not enough to say, with Plato, that theoretical objects—*forms* or *ideas*, call them what you will—are discovered in the context of inquiry or research. It is necessary to explain how this occurs.

Thirdly, scientists set themselves the goal of discovering something that is general, that is, of arriving at a result that is not limited to "configurations in progress." The goal is to arrive at a single formulation which will account for all similar cases. Consequently, the peculiarities of individual objects do not concern them. This explains the practice of selecting specimens for research which are, so far as possible, free from confounding factors. Thus the water used in research will not be from the local river, but distilled water, the chemicals 99.9% pure, the population surveyed selected with regard to specific criteria, etc. The rationale for this selection is that a researcher wishes to discover something that remains the same in all instances of the kind of thing being studied. To isolate such a factor requires excluding the peculiarities of particular cases, so far as this is possible. In scientific research one is interested in a particular specimen only in so far as it is representative of an indefinitely large range of instances of the same thing. In other words, scientists are interested in finding a theoretical object that will account for an indefinite number of individuals. The significance of the theoretical object is that it can be used to identify an unlimited number of instances of the same kind of thing, an indefinite number of possibilities in other words. The goal of reaching an object that identifies a class or set of things seems to be a fundamental part of all scientific work. Does this mean that scientists themselves are essentialists? Other than pointing to the context of practices, Ginev does not clarify why this is not the case.

A fourth concern is that the reflection on the status of science's theoretical objects is like the problem of the hermeneutic circle. In a rather turgid sentence we are told that "the status of science's theoretical objects has to be revealed by the same transcendental reflection that unfolds the hermeneutic circle between the fore-structure of interpretation and explicit structure," because "the very circularity is mediated by the interrelated practices of normal scientific research" (Ginev, 2009, p. 383). The key question is this: What is the meaning of "mediated?" The parallel may be instructive, but now one faces dealing also with the problem of the hermeneutic circle.

Finally, a more general comment. While Ginev's account of the constitution of meaning is nuanced, noting the changing configurations characteristic of even normal science, and again the progressive articulation or appropriation of possibilities, nevertheless one is left with the impression that this account of science is from an external point of view, an account of it as a phenomenon spread out in space and time, and in terms of results, rather than seen as arising though some inner dynamic found in each and every scientist, and whose shared questions are the basis for cooperation in a field. No doubt there are references to scientific research being a mode of being in the world of Dasein, many references to "self-referentiality," and even mention of "the interpretative nature of human beings" (Ginev 2009a), but what this interpretative nature is and how it operates is not explored.

3. An Alternative Account of Theoretical Objects

Ginev asserts that theoretical objects arise in the context of practices. While true so far as it goes, this is not saying enough. It is more accurate to say that theoretical objects arise in the context of inquiry. To inquire is to ask a question, to ask *what?* or *why?* or *how?* Inquiry is an activity of an inquirer, of a subject; it is a mode of being of Dasein. Also, a question is a response to some experience, something which one has seen, heard, or noted in some way. The "ready-to-hand," to borrow a phrase from Heidegger, ceases to be just familiar, but also becomes strange, or puzzling. What was merely there as a familiar object is now "seen afresh" and recognized to have aspects or elements which have not previously been noted. In short, we begin to wonder, and if time and energy allow, inquiry follows.

Beyond the genesis of inquiry, it should be added immediately that we never just question, but that our questions arise in a context. Certainly this is true in the with regard to the present discussion, scientific research. And so, it must be added immediately that we do not just inquire, but we inquire with certain presuppositions or expectations, what Bernard Lonergan has called "heuristic structure" (Lonergan, 1992, Ch. 2, "The Heuristic Structures of Empirical Method," pp. 57–90). Modern science has a long tradition that is unified by such expectations. One illustration will suffice. In the context of scientific research, we investigate only hypotheses which have some kind of sensible consequence which can be observed. Hence, a hypothesis which has no observable consequence is excluded from science. In other words, if there is no way to test an idea or hypothesis, then it cannot be a part of empirical science, for there is no possible relevant research. To put the point in Ginev's language, there would be no empirical practice to relate to such a theoretical object. Hence, theoretical objects with no testable consequences are excluded from empirical science. A second, related point is that it is not sufficient for a hypothesis to have empirical consequences, but these consequences must be confirmed in scientific research. Those whose consequences are not confirmed by observation are discarded (Cf. Lonergan, 1957, pp. 95–96).

The above two observations about theoretical objects illustrate the contextual nature of all theory. A theoretical object does not appear without prior conditions. There must be some event or thing, and a person who is alert enough to note, to identify, to attend to an aspect of familiar experience which she has not yet understood, *wonder* as Aristotle called it. So two conditions must be met for research to occur: without the experience or a thing there would be nothing to inquire about; and without the person being intellectually aware, being curious, the experienced would simply remain a familiar object, something ready-to-hand.

One might conclude from the reference to experience that practice is prior to theory, but this would be a mistake. Some kind of experience is prior, for without this there would be no occasion to begin to puzzle, to wonder, and it is wonder that is the genesis of the process of inquiry. And when we wonder we naturally seek and try to discover, to grasp a possible correlation among the elements in the experience. And we follow up that possibility by formulating what we have discovered. Discovery leads to the formulation of hypotheses which are composed of theoretical objects.

What is the goal of inquiry? One who inquires seeks to understand "something as something," to grasp the already familiar in a new way. Or one might say, the goal is to give an alternative account of the thing that is already familiar. Thus a physicist gives an account of light not in terms of color but of wave length, etc. But to give an account one must articulate one's grasp of some possibly relevant correlation. And so it is that a hypothesis and theoretical objects generally are a response to inquiry. A hypothesis is generally the identification of some correlation that has the potential to be relevant or significant—significance and relevance being understood in relation to the question that has been raised.

This aspect of the matter needs some elaboration. The significant is what is relevant, as opposed to what is incidental. And it should be added that significance is determined by the question that has been asked. For in any situation, there are multiple questions which might be asked, and depending on the question different aspects of the experienced object are relevant. To borrow an example from ordinary experience, by virtue of their habits and interests a logger and an ornithologist walking into a forest will observe very different things and so inquire very differently into what is present in the environment. Each will observe some elements and equally will ignore others.

If it is true that a hypothesis is formulated only in the context of experience, the fact is that formation of a hypothesis (theoretical objects) is just one stage of the cognitional process. What a scientific researcher seeks is to determine whether the hypothesis can be applied to all similar situations. This is what we mean by verification. It is assumed that the particularities of a situation, place and time for example, can be ignored. So, for example, whether I run a given test in Bowling Green or in New York is regarded as incidental, as irrelevant. Equally, whether I run the test or someone else is also incidental, so long as both of us follow appropriate procedures. What each scientific researcher is seeking is to describe are the significant differences. And what constitutes a significant difference is determined by one's hypothesis.

Another element of theoretical objects as they are found in hypotheses must be highlighted. They are provisional. When one speaks only of "theoretical objects" this aspect is obscured, but the very term "hypothesis" indicates that its content is to be tested, that it is a provisional formulation needing to be confirmed by appeal to experience. Note that there is a double relationship. It has already been said that without experience, there would be no inquiry, and so prior to the formulation of a hypothesis some experience is required, and this experience rouses one to seek to understand. But now it must be added that formulating a hypothesis is also not enough. Reflection follows. A hypothesis is the formulation of an insight; it is the grasp of some possible explanation. Reflection by contrast asks: "Is this what is going on?"—*this* referring to the content of the hypothesis. And how is this question answered? By a return to the data, by checking whether the hypothesis accounts for the data, or whether there remain relevant aspects unaccounted for. If there is a remainder, one revises. Only when all relevant aspects of the data have been accounted for does one affirm the hypothesis. Theoretical objects are initially hypotheses that, once formulated, require confirmation which is accomplished by a return to the data.

Even this way of describing the provisional character of theoretical objects is not sufficient, for the process of checking whether one's hypothesis is adequate is repetitive and progressive. No sooner has a researcher confirmed a correlation in a study, than she is likely, given the opportunity, to test the same hypothesis in a slightly different situation, varying the object or the procedure in some manner. Thus a method successful in one area may be extended to another, and even eliminate a long familiar approach to problem, as Ginev illustrates from the study of the phenomenon of florescence (Ginev, 2009, pp. 378–80). In sum, a hypothesis may be adequate in a particular context, but the larger field in which it stands may come to be questioned, and even discarded.

Enough has been said, perhaps, to set forth the implications of this discussion for the discussion concerning the status of theoretical objects. (It seems more natural to speak of hypotheses, explanations, or correlations, but since Ginev formulates the problem in terms of theoretical objects, I will use this language here. Since hypotheses and explanations are combinations of theoretical objects, the relevant points can be made using either formulation.) A first point to be made is that theoretical entities as they have been described are not *essences* in the sense that Ginev rejects—"something independent of the dynamics of practices of normal scientific research" (Ginev, 2009, pp. 382). As noted they arise in the context of inquiry, are a result questioning. They are a product of a person thinking about something; they are experienced as a discovery, even it is recognized as a provisional discovery which will be open to revision or even reversal.

The last point needs more elaboration. It is recognized in many, if not most, cases that a hypothesis, even when confirmed, is provisional. Clarification may be needed; more precision in measurement required; discovery of an unexpected complication may require reformulating what constitutes the relevant factors, and in consequence a revision in one's understanding of what dimensions of the object are relevant or irrelevant. Because of these kinds of issues, a field of inquiry develops and expands. Later researchers build on the work of earlier researchers, and so gradually a body of knowledge results.

The mention of cooperation among researches raises a new issue. Two points need to be made with regard to this development. Development results from collaboration and yet is advanced by individuals. The possibility for these two dimensions must be noted. Every hypothesis is an identification, or better, specification of some set of conditions. Verification of a hypothesis is a confirmation that the specified conditions are actually found in experience.

Both formulation and verification are activities carried out by researchers; they are products of the development of understanding in individual persons—a mode of Dasein's being in the world. How is it then that what is discovered is not a private possession? In so far as a hypothesis is a formulation of a certain way of understanding a situation, it is a specification of criteria for its being affirmed. Both the criteria and the conditions are open to others. The criteria are grasped by anyone who understands the hypothesis, for to understand a hypothesis or correlation consists precisely in grasping its empirical conditions. Secondly, for anyone with competence in the field, anyone familiar with the relevant practices, the conditions are available for inspection. Hence there is the possibility of grasping to what extent the predicted conditions actually pertain. While verification is a judgment which is a personal achievement accomplished by each and every researcher, the result of such verification is open to every other person who grasps the meaning of the hypothesis and has occasion to examine the relevant experimental evidence.

Implicit in the above account is an anti-essentialist account of theoretical entities. There are a number of features of this account which are anti-essentialist. First, according to this account theoretical objects are the product of inquiry. They are the result of a researcher responding to a particular object or event by grasping some correlation that appears significant and then formulating what is significant or relevant. So it is the product of a "thinking out," an expression by a researcher about what is important or significant in an experience, or body of data, and by implication what can be ignored, because it is incidental to the inquiry at hand. Typically, as was noted, one thing which is incidental is the "here and now" character of each experimental object. What one is interested in is not that some X occurred at some place or time but some difference in what occurred at the particular place and time. To put the point in another way, a physicist studying the oxygen atom or a chemist studying water is not concerned with the specific atoms or molecules selected for study, but only that the sample is free from impurities. The incidental character of time and place makes it possible to generalize from a small experimental base. In addition, it makes scientific collaboration possible, since within limits, time and place are irrelevant. This account is a description of what Ginev calls "meaning constitution."

Theoretical objects are so significant, because once one has grasped a meaning, then every other situation which is the same in the relevant respects can be understood in the same way. No new insight and formulation is necessary, and after a few repetitions, what we have previously understood soon becomes obvious and familiar. Once this is the case, typically we are able to attend to new matters. To illustrate, having learned letters, we can attend to words, and having mastered words we can focus on sentences, etc. It is also the basis for what Ginev characterizes as the everydayness of scientific research. Yesterday's discovery becomes today's assumption.

Secondly, the provisional nature of theoretical objects has been noted. This point strikes directly against the essentialist conception of theoretical objects. If theoretical objects are a product of a person "thinking out" in response to experience, then they are also merely a possibility, a thought which is being entertained by someone. The key aspect is that inquiry does not stop here. It is not enough to come up with a theoretical object, the whole point of formulating a hypothesis is to be able to test it. Testing or verification is a response to a different requirement of mind. Every idea or theoretical object or hypothesis must be confirmed. If not confirmed, it is only a possibility that may turn out to be "a bright idea," speculation, a fiction, etc.

One might say that the first stage of scientific inquiry results in the formulation of a theoretical object or hypothesis, an answer to the question *what? why?* or *how?* that can be tested. Answering this question is not the end of inquiry, but leads directly to a different concern, the critical question, *Is it so*? *It* here refers to one's formulation, the content of one's hypothesis. Verification is one name for this second question and resulting process. For it is a process, another kind of "thinking out." If the first result of inquiring into experience is discovery of some correlation which is fixed by means of formulating a hypothesis, then the result of having formulated a hypothesis is that a new question arises: Is the proposed understanding, the hypothesis, correct? From trying to understand one moves to determining whether one's understanding is correct. As naturally as experience gives rise to inquiry, so in inquiring we shift from trying to understand, answer the question, *What is it?* to *Is it so?* For inquiring is a complex process, where we are not satisfied with acquiring some understanding but we also want to know whether our understanding is correct.

How is this new question answered? Either by *yes*, or *no*, or by *perhaps, possibly, probably*, and the like. We have a large number ways in which we indicate that no definitive answer is possible. What is going on in verification? To name the process is not enough, especially since so much puzzlement results from not being clear on this element of knowing. How do we answer the question, *Is it so?* In other words, when is it permissible to affirm a hypothesis? Here it is relevant to begin by noting that a hypothesis arises as a result of some grasp, some insight, into a particular set of conditions. Hypotheses have their origin in experience and in an inquiring mind. A hypothesis is the formulation of what is significant in the experience, of some relevant correlation in it, etc. Now, having formulated what is relevant, we turn to checking, by returning to the data to determine whether it is really so, whether the specified conditions are present in experience. (There is, of course, also the question whether the hypothesis has been formulated correctly, that is consistency in terminology, etc. In other words, logical concerns must also be met, but I am not going to elaborate on this aspect here.)

A hypothesis is a formulation of a set of conditions. As already noted above, a hypothesis can be part of empirical science only if the conditions specified can be detected in some kind of experimental setting. (The critical phase in knowing is equally apparent in everyday living, as when we question the claims of others: "You said such and such but are you sure it is so?")

Verification, then, consists of raising the critical question, *Is it so?* But what is the basis for an answer? Recognizing that every hypothesis is a statement of conditions supplies the clue to an answer. A hypothesis is a conditioned; it can be affirmed only if its conditions are fulfilled. And so it is that after having formulated our question, we answer the critical issue by returning to the data from on which our hypothesis was based. Formulate a hypothesis about water and one will return to water to determine whether the specified conditions are fulfilled. How, then, are we to understand the way in which the critical question, *Is it so?,* is answered? An initial observation is that we affirm and deny or express some modality of probability only after we have reflected. To reflect is to survey the data with regard to whether it reveals the conditions which are specified in the hypothesis which has been formulated. Negatively, reflection is not a new attempt at understanding, but rather a movement of mind in which one asks whether the conditions which have been specified are manifested in the relevant data.

At one level this answer is correct, but at another level it is so incomplete that it is likely to be misunderstood. A preliminary clarification is found by observing the actions of a researcher. After coming to a conclusion that something specified conditions have been observed and all ambiguity removed, there is no further point to repeating the operation and to do so is even considered a sign of stupidity. Thus well-established findings in a field need not be repeated. Rather having satisfied ourselves on a particular point, we turn to a new issue. And so having discovered that salt can be analyzed into chemical elements, one might turn to apply a similar technique to another crystalline substance, etc. So it is that Galileo's measurements of acceleration on an inclined plane could be extended to freely falling bodies, etc. And so after verifying a hypothesis in one instance, one may turn to replicate the results in a new situation or apply similar techniques to other materials, etc. And so it is that individual judgments that flow from answering the critical question are not just individual occurrences, but stand in complex relationships to one another constituting what we call fields of knowledge.

A more satisfactory account must explore some of the complexities flowing from various possible answers. Two possibilities exist – that the data conforms to expectations or that it does not. The second case is the easier one: if the conditions specified by the hypothesis are not found in the data, then the hypothesis is discarded. In the first case the response is more complex. In the simplest case a hypothesis relates to a limited body of data which can be inspected exhaustively, and then a conclusive answer can be given. Ask "Are you in your office?" and the answer requires only that one be aware of where one is. Admittedly, doubts could be raised about whether one is deluded, etc., but note that this would result from raising other questions and not really be an answer to the question in the context in which it was asked.

There is, however, a more complex case that is typical of an affirmative answer to the critical question—that only a part of all possible relevant data has been examined. Then, the affirmative answer brings one sequence in knowing to a conclusion, but inasmuch as additional data is available, the process can be repeated, indeed almost begs for new inquiry to be undertaken. And so it is that each individual judgment or verification stands in the context of others. The fact that each individual judgment stands in the context of many other judgments has important implications, a few of which need to be mentioned here. In each field of knowledge the relationships among judgments differs, and so here I will note only a couple of characteristics as found in the empirical sciences.

In the empirical sciences the answer to one question assumes the answer to a multitude of previous questions. So complex previous ways of understanding a particular phenomenon are assumed and it is only to the extent that such understanding is assumed to be correct that one can go one to make a new affirmation in the present context. For example, to talk about wave length, one needs to assume a theory and practice of measuring. A conclusion, then, may be affirmed as long as the assumed system of measurement is accepted, but in fact systems of measurement are subject to revision. New technology may allow more precision or on the theoretical side a new theory may require correcting for what was before thought to be incidental. Enough has been said, perhaps, to make the point that an affirmative judgment with regard to a hypothesis may be regarded as merely probable, because in fact the hypothesis being considered stands in the context of a much larger set of conditions. So it is that even what is verified may be regarded as merely probable insofar as it is dependent on techniques and theories that are open to revision.

Prof. Ginev has argued for the immanent transcendence of theoretical objects. I have preferred to frame the discussion in terms of hypotheses, but since hypotheses are made up of theoretical objects, what applies to hypothesis is applies to theoretical objects. In the above analysis I have tried to elaborate on "the immanent transcendence" of theoretical objects, for a theoretical object is a single content which has the potential to be applied to an unlimited number of instances. That potential range is often characterized as a *horizon*, or in classical language a *universal*, a single content which has the potential to be applied to an indefinite number of situations of a specified kind. Something more needs to be said with regard to how this account avoids falling into essentialism, but this is best treated in a separate section.

4. Beyond Essentialism

In the previous section I have given an alternative account of the "immanent transcendence" of theoretical objects, focusing on locating theoretical objects in the context of inquiry rather than in terms of practices. The implications of this shift have been only partially examined. Also, additional comment is needed to clarify some of the assumptions in this alternative account. Contrasting these assumptions with those embedded in essentialism will also aid in clarifying how this account differs from Ginev's account where he does not break completely with essentialism.

First, theory has been set in the context of inquiry rather than of practices, as Ginev prefers. Without doubt there is a close relationship between theoretical objects and practices. Ginev is right to note the "changing configurations of scientific practices" (Ginev, 2006, pp. 110–123). Practices gain their significance in relation to one's theoretical framework, and so their status depends on the theory which they exemplify. The co-dependence, as it is described, is certainly the case. But it is an asymmetrical relationship. The theory gives significance to the data, but the data supplies the conditions so that the theory may be affirmed. In this regard there is a kind of priority which accrues to theory, for theory specifies what becomes relevant and it is in relation to theory that the data of experience provide confirmation or disconfirmation, so that what was entertained as a hypothesis is either affirmed or denied.

There is another role for experience, but it is prior to theory. Experience is a source of meaning. It is the material on which intelligence works or with which it works both so that questions arise and as that which is examined when one is questioning. Understanding is not a free creation, but rather theoretical entities or ideas are a response to the materials supplied in experience. Since experience is variegated and variously significant in different respects, different ranges of experience are relevant for different theories. (The opposite situation where idea is prior and instantiation of the idea follows is found in the work of an artist, who works to bring what was only a sketch or suggestion into existence.)

This role of experience as source of meaning precedes and complements the other, and reveals that theory is both a response to experience and confirmed by experience—a double relationship exists. This fact supports the view that inquiry is a process, an ordered process, in which each of the parts must occur in the proper order. Without experience, the presence of data, there is nothing to understand; and experience alone is without understanding, like Aristotle's man of experience who may act effectively, but is unable to explain. Similarly, understanding by itself is not enough, for by itself it is merely a possibility, a possible correlation. What is desired is not just an idea, but the confirmation that what one has conceived is adequate to experience. So, understanding is naturally followed by reflecting and judging. Finally, as already noted, as human beings we attend to one thing at a time, and every judgment is a limited account of all that can be known, and so our single judgments stand with others both before and after and so coalesce into fields of inquiry.

To focus on inquiry rather than practices in an account of theoretical objects is significant because it places them in a larger context, in the full sweep of our experience of knowing. It opens the door to a fuller exploration of the dynamic of our quest to understand.

Another way to get at the differences between the above account of inquiry and cognitive existentialism is to observe that Prof. Ginev is developing his philosophy of science in context of the program which Heidegger outlined in *Being and Time.* In this conception the *ready-to-hand* is objectified in the *present-at-hand*: through the process of interpretation meaning is constituted. For Heidegger there are three moments in this process. Ginev describes them in this way: there is "meaning we have in advance (a fore-having of possibilities), meaning we see in advance (a fore-sight of possibilities), and meaning we grasp in advance (a fore-conception of possibilities)" (Ginev, 2009, p. 381). Each of these requires clarification.

"Interpretation," writes Heidegger, "is grounded in *something we have in advance* – in a *fore-having*". In elaboration Heidegger states that "the interpretation operates in Being towards a totality of involvements which is already understood" (Heidegger, 1962, p. 191). *Fore-having* appears to be an account of the ordinary, everyday world that is familiar to us, the lived world, the world of common sense, however one wishes to identify it. The point is that there is given in experience that which can become a source of meaning; the empirical is something which potentially can be understood. Since it is what precedes interpretation, it provides the materials about which an inquirer can come to question. It is another way of identifying experience as a source of meaning.

Just as the response to inquiry is the formulation of some hypothesis (theoretical objects), so "interpretation is said to be grounded in *something we see in advance* – in a *fore-sight*". According to Heidegger, *fore-sight* "takes the first cut" out of what has been taken into our fore-having, and does so with a view to a definite way in which this can be interpreted". What Heidegger seems to be describing here is the fact that when we inquire, we formulate our discoveries—that "towards which we set our sights 'fore-sightedly', becomes conceptualizable". Again there are notable similarities between Heidegger's account and one focused on inquiry. All inquiry is, as I have noted, in terms of a particular question and so can be said to be "with a view to a definite way." In addition, formulation is an identification of what is relevant, significant, or essential to the insight which has been gained; this is what is "conceptualizable." However, if some aspects of experience are relevant, it is also the case that others are irrelevant, at least for the moment, in relation to the question which is being asked. In this way inquiry also "takes a first cut" out of what is present in the *fore-having* (*ibid.*).

In addition, this meaning is rightly said to be "something we see in advance," for every theoretical object is an articulation of a content which is potentially found in an indefinitely large number, if not infinite number, of situations. The point is that once one has made an interpretation in one context – experimental setting, in the case of scientific research – the same understanding has the potential to be employed in other situations similar in the relevant ways. So it is that theoretical objects can be described as something "seen in advance." More precisely, their openness to repeated employment is cast into language which describes the range of possible instantiations as a horizon. The visual metaphor implicit in this account is satisfactory if it is understood that what is being described is the fact that a single understanding, one theoretical object, has the potential to be employed in an indefinite range of instances. In other words, it is, as found in the mind, a universal.

Finally, fore-conception is said to be "*something we grasp in advance*". In clarification of this dimension of meaning constitution, Heidegger states that "the way in which the entity we are interpreting is to be conceived can be drawn from the entity itself, or the interpretation can force the entity into concepts to which it is opposed in its manner of Being". Here, Heidegger's account of the constitution of meaning diverges significantly from the account of inquiry given above. No doubt Heidegger is right to observe that interpretation can be drawn from the entity itself or it can force an entity into concepts that are opposed to its being, but what the reader wants to know is this: how do I determine which one is occurring in each particular situation? Am I giving an account of what is present in fore-having or am I misconstruing its nature? Heidegger is raising the critical question. This is clear from the fact that he speaks of the interpretation being decided "either with finality or with reservations" – with certainty or merely some modality of probability. On what basis are interpretations decided with finality or with reservations? Nothing is said at this point. What is lacking is any reference to reflective understanding and its product, judgment (*ibid.*).

Here lies the most significant difference between Heidegger and the account of inquiry sketched above. Without an account of the critical dimension of understanding one raises the issue of the hermeneutic circle and the issue appears insoluble. Here it is not possible to do justice to this complex issue, but an elaboration of the implications of the above account of inquiry can indicate where the solution lies. How do we know when we have arrived at an interpretation which is "drawn from the entity itself" and is not "forcing an entity into concepts opposed to its being?" The key is to realize the progressive nature of the process of learning. There is no doubt that my current understanding stands in relation to other concepts and supporting evidence, such that the part gains its meaning in the context of the whole and the whole gains its meaning from the parts. This mutual dependence or circularity, however, is not that of a closed or vicious circle. The reason is that individual judgments stand in the context of other judgments based on related aspects of experience. Having examined one area of experience, the intellectually aware researcher will naturally seek further evidence from other related ranges of experience. So long as the theory being employed is adequate to explain the new ranges of data, no revision is required; however, if some aspect of the data cannot be explained using the old theoretical framework, then a new account will be necessary. So it is that theoretical objects are progressively redefined even in the course of normal science, and a major shift in one's scientific theory occurs when a new way of organizing the data in a field is the only way to account for all the data. Again, think of Ginev's account of the history of the study of florescence serves as an example of the latter phenomenon. In other words, the process of learning involves successive judgments, each of which stands in the context of others. In the case of empirical science all judgments remain provisional, because there will always be the possibility of additional data or new ways of grasping the data of experience. (For more details, see Lonergan's analysis of probable judgments, *Insight*, pp. 324–329).

Another implication of the account of reflective understanding and judgment relates to the gap that Ginev identifies between the ontical and the ontological in Heidegger, roughly the results of specialized inquiry as opposed to being itself. According to the account of inquiry given above, scientific research proceeds to answer the question, *Is it so?* An affirmative answer, *yes*, is an affirmation of being, not merely some entity; or if one prefers, the theoretical objects, things, or entities that are affirmed in science are a part or aspect of being. In so far as a scientific question is only one of a number of ways in which we can question our experience, the answer is limited to an expression of one dimension of being. Other questions asked in different ways with different criteria abound. One only need note our common, everyday way of dealing with particular things for a relevant example. And so it is that science is only one way of grasping what is, for other questions with different criteria are possible. Our desire to know, our capacity for inquiry, is not limited to scientific research; rather scientific research – admittedly an enormously influential question in modern and contemporary times – stands in the context of the many questions that we human beings struggle with both individually and collectively. Indeed, all science ultimately has its source in everyday experiences and its discoveries result in applications which equally find their place in the everyday world of common sense.

As noted in the beginning, Ginev's account of science is both subtle and detailed. The primary way in which the account of inquiry which has been outlined above differs from his view is in sketching the complexity of human knowing, that we proceed from asking what something is to reflecting on what we think we have discovered to determining whether *it is so*. As experience leads to discovery and is followed by the formulation of understanding, so understanding leads to asking whether our understanding is correct. Recognition that every hypothesis is a formulation of conditions, and every judgment an affirmation that the conditions specified in the hypothesis are present in experience, clarifies the respective roles of theory and data. To be a part of science, theoretical objects (terms) must have empirical consequences; determining whether these consequences are actually present in experience is the task of reflection, the critical question. For every hypothesis is a formulation of conditions, and unless those conditions are found in experience, the hypothesis cannot be affirmed. For Ginev the fore-structure of interpretation and the explicit structure of science are "mediated by interrelated practices" (Ginev, 2006, pp. 135–144). While this account includes a reference to both theory and practice, it fails to articulate the dynamic clearly. The contextual nature of theoretical objects is reflected in this account, but their provisional character is not recognized. The provisional nature of theoretical entities becomes clear only when one recognizes that beyond understanding, beyond grasping *what is*, there lies the further question, *Is it so?* Without this clarification, there is the tendency to regard theoretical objects as essences. With this clarification both their generality and their provisional status has been described. And so the tendency to essentialism has been accounted for and corrected.

In addition, when the role of reflection and judgment are noted, then the problem of the hermeneutic circle is replaced by the process of learning rooted in our desire to know all that beckons in experience. Experience rouses inquiry, and inquiry seeks understanding, and provisional understanding seeks confirmation, grounding of the understanding. And far from ending there, once satisfied on one point, we turn to inquire about a related issue or new data in experience which has the potential to call our hypotheses into question and so trigger the quest for a more adequate understanding. And so it is not appropriate to think of inquiry as a circle moving from experience to understanding and back to experience, but rather as an ever expanding quest for a better understanding grounded in ever greater ranges of data. In this framework essentialism is no longer a concern, for the provisional nature of all theoretical objects is clearly recognized and the generality of theoretical remains. The immanence and transcendence of scientific thought are both recognized and accounted for.

WORKS CITED

Ginev, Dimitri. (2006), The Context of Constitution. Beyond the Edge of Justification, (Boston Studies in the Philosophy of Science), Springer, 2006.

————. (2008), "Cognitive Existentialism", *Iyyun (The Jerusalem Philosophical Quarterly)*, July 2008, pp. 227–242.

————. (2009). "From Existential Conception of Science to Hermeneutic Phenomenology of Scientific Research." *Journal of Philosophical Research*, Vol. 34: 365–389.

————. (2009a) "Perspectives on Reality and 'The World' in Scientific Realism Debates", *Human Affairs*, 12, No. 2.

Heidegger, Martin. (1962). *Being and Time*. Tr. by John Macquarrie and Edward Robinson. New York: Harper & Row.

Lonergan, Bernard. (1957, 1992). *Insight*. Toronto: University of Toronto Press, 1992.

Plato. (2002) *Meno* in *Five Dialogues*. Tr. by G.M.A. Grube; revised by John M. Cooper. Indianapolis, IN.

Two

THEORY AND THEORETICAL OBJECTS IN AN EXISTENTIAL/HERMENEUTIC CONCEPTION OF SCIENCE

Robert Crease

An existential/hermeneutic conception of science recognizes that science is only one way in which human beings engage the world, and not the default setting, either. Human beings connect with the world through a matrix of practical attachments – phenomenologists call it the *lifeworld* — that precedes any cognitive understanding. Human beings, moreover, are not automatic information absorbers, but are productive and sense-making; they have an *interpretive* or *hermeneutic* relation to the world and to themselves. While much traditional philosophy of science is thin or anorexic and does not flesh out all of the ways in which science is an interpretive, sense-making activity, or how this activity is bodied forth in the world, it is the merit of the existential/hermeneutic philosophy of science to trace how the interpretive, sense-making activity of science arises out of the lifeworld. The aim of this paper is to discuss the nature of theory and theoretical objects in such a conception, using Dimitri Ginev's work and a recent critique as a starting point.

1 Dangers of Theorizing about Theory

It will be helpful to begin by reviewing several dangers that beset philosophers who speak about the nature of theory and of theoretical objects in science. A first is to begin by speaking about a particular kind of theory and using that as the emblematic instance, assuming that what is true of it is true of other kinds of theory as well. Theories, after all, are of many kinds and play many roles. Are we talking about mathematically structured theories in physics such as quantum electrodynamics? Are we talking about biological theories such as Darwinian evolution? What about what we mean when we speak of things like "theorizing race?" And of such theories at which stage? A philosophical approach has to beware of knowing ahead of time what something is before inquiring into what it is.

A second danger is to prematurely formalize the process. Many accounts of theory development for instance, looking backwards, reconstruct it as a simple, step-by-step process – "First we hypothesize, then we confirm" – and then claim, "This is how it must have been done!" That's somewhat like reconstructing a child's acquisition of language by saying that the child must have learned language by acquiring one word after another, or reconstructing perception by saying it consists of the apprehension of sense data, then its organization. A philosophical approach has to beware of knowing ahead of time how something works before inquiring into how it works.

Yet a third danger is present in the form of a tradition that tells us what we have to say about something – for instance, that knowledge consists of representation, that the acquisition of knowledge is the acquisition of representations, and that theory is a particular kind of representation. If we begin this way, we already saddle ourselves with insoluble problems. Moreover, we prevent ourselves from being able to connect with scientists; theory and theoretical terms are one place where the experience of scientists, and even students who work in labs, collides with "the story," with what traditional philosophy of science tells scientists how they work and what they do. Small wonder they get incredulous and dismissive of philosophy of science.

An existential or hermeneutical starting point recognizes that scientific practice consists of ongoing inquiry within a changing situation. Each scientist is historically situated within a particular historical, concept-laden discourse and within a particular historical, technological environment, interpreted according to available historical resources. Moreover, in scientific practice it is not a question of a solitary cognitive subject confronting a specific object or problem, nor a collection of objects and problems, but rather of a historical community of living, practically engaged researchers confronting a historical, holistic situation. Even in situations of apparent solitary breakthroughs – Einstein pursuing general relativity in his study, Heisenberg pursuing matrix mechanics on Heligoland – such thinkers were informed by and responding to a community of fellow researchers and a situation. Moreover, t*his* generation of researchers in a given field is not only a different community but faces a different situation from the previous generation, and the next.

All this is simple to state, and fairly obvious once one thinks about it, but is easy to miss. The products of science history are profounding distracting. How compelling it is to focus on the products of science, or its theories, rather than on the self-creating process to which products and theories belong! Genuine science involves a force, a pressure that extends outwards to the future. Seeing the vestiges of that force in the past does not necessarily help us to understand it in the present. We can call that force curiosity, ambition, desire, and people will understand what we are saying, but this can easily result in a static conception of science. We can edit out this force from our picture of science, but what we see then is only its remainders. Science is a process by which new forms, new concepts are created; growth and change, a movement of differentiation, is integral to it. A true philosophy of science has to not just *allow* for there to be fresh creations, new forms, continuous enrichment of the lifeworld, but to show the drive, pressure by which these are produced as a natural movement in inquiry.

It is refreshing to encounter Ginev's work, for he is engaged in an existential/hermeneutical conception of science, is aware of the dangers, and strives to avoid them. In the following discussion of theory and theoretical objects, I shall first summarize relevant parts of his work, then use a recent critique to elaborate my own thoughts about theory and theoretical objects (Ginev, 2011).

2 Ginev's Cognitive Existentialism

In *The Tenets of Cognitive Existentialism* and elsewhere, Ginev announces a two-step revision of the existential conception of science inherited principally from Heidegger. The first step is to place the practices of science in terms of everydayness; research in progress involves much "everydayness," not just mathematical theorizing. The second is to show that mathematical projection is only one way of doing research. These are, I think, the right moves, and together they constitute what Ginev calls "cognitive existentialism". Science, Ginev thinks, does not so much "de-world" as "re-world", shifting "the horizons of projected possibilities". Scientific domains are "disclosed within a characteristic hermeneutic situation" that can persist across interrelations of practice. "Characteristic hermeneutic situation" then becomes the key concept of cognitive existentialism.

At the beginning of *Tenets*, Ginev announces his ambition "to develop a picture of scientific research as a process of appropriating projected possibilities within the interrelatedness of practices of inquiry." This will involve advancing a view about "the status of science's theoretical objects," which he calls "inscriptions" on a horizon of possibilities for doing research, artifacts of the various practices of science. He describes this, appropriately, as a realism but specifies that it is a "hermeneutic realism," a term that has been used before by others.

The motive for this work is said to be as follows. In dealing with theoretical entitles, hermeneutic phenomenology of science has viewed theoretical objects in a dual perspective: in their relation to (constituted by) other parts of the theory, and as related to (constituted by) what Ginev calls "normal scientific everydayness," or its practices and intentions and so forth. The first has a structural lawfulness and, Ginev fears, courts essentialism. The other amounts to a horizon-projection, and does not run that danger. Hermeneutical phenomenology of science is then faced with the task of characterizing each, and describing their relation.

Ginev refers to "two large hermeneutic circles" in the process. One is the way that theoretical entities, constituted by relation to each other, are "hermeneutic instruments" that supply the background for interpretation of research, opening up domains for exploration, pointing to key problems and likely solutions, and so on. The second circle is the way that that theoretical entities arise out of the "everydayness of scientific research," its concerns and practices; the key phrase is "objectifying thematization." Their relation seems a puzzle, a puzzle that Ginev captures in the phrase "immanent transcendence." What is the relation between what he later calls research objects, which you have in the lab, and theoretical objects? What is the line between them?

It's a puzzle, because the theoretical one appears to reach beyond the present context and practices – to be not exhausted by the way it appears in the here and now. It looks like it has a "cognitive essence," and that what scientists' practices do is to "operationalize" these theoretical objects, these invariant objects, by "transforming them into research objects that are ready to hand and present at hand in normal science" Ginev says this is opposite of his view.

Ginev uses Kockelmans's work as a foil to work out some of the main characteristics. Again, it involves a twofold conception, of an "open horizon of mathematical idealizations" and a "constitution of relevant empirical facts in the practices of scientific research." It involves an "ontological projection of a scientific domain," and thus has a transcendental dimension. Ginev also uses van Fraassen's constructive empiricism as a foil in another way, to show the potentials and pitfalls of avoiding reifying theoretical objects. An example involving molecular biology is helpful in elaborating this.

Ginev then refers to the "central problem of hermeneutic phenomenology" as describing the interplay of the world as the organized equipments and practices (intramundane reality) into which we are thrown, and the world as the horizon of possibilities whose "interpretative appropriation articulates" the world in the sense of intramundane reality. In this course of this, he makes an interesting critique of Brandom and his neo-pragmetist approach to the existential analytic, and uses Misch's critique of Heidegger as a foil.

Cognitive existentialism, in short, "rejects the view that scientific research is 'just looking, without circumspection.'" Heidegger is wrong to say this, as he does in *Being and Time*. Rather, Ginev says, "the point of departure of cognitive existentialism" is that "scientific research constitutes its own circumspection." "It does not de-contextualize bur rather re-contextualizes what is constituted within the interrelatedness of scientific practices." The real critique of Heidegger, the central one, it turns out later, is that Heidegger's insistence that research is preceded by and opened up by a "mathematical projection" that is independent of and prior to the rest of science is incorrect. Rather, Ginev will say, this research horizon is opened up and disclosed "by the interrelatedness of all practices taking part in its constitution"

Cognitive existentialism, therefore, "defends this autonomy [of scientific research] without presuming any kind of epistemological justification of science's authority … [it] champions the ethos of scientific research without succumbing to scientism."

Chapter Three describes how this happens, via the concept of a "characteristic hermeneutic situation," or the unity of the three elements of the hermeneutical circle "as embedded in a domain's interrelatedness of practices." But it begins with a welcome illustration, from theory of optics. The point is to show that a domain of research is opened up not by a "mathematical projection" that is independent of and prior to the research, but by "the interrelatedness of all practices taking part in its constitution." Section 2 of this chapter is about the notion of a "thematizing project," by which he means a "macro-historical reproduction of a characteristic hermeneutic situation that stretches beyond the development of particular scientific domains, exhibiting thereby stable (historical) types of scientific research." Ginev distinguishes five principal thematizing projects in biology, which are ideal types on a spectrum stretching from a pole of taxonomic objects to a pole where the objects require maximum "hermeneutic preunderstanding". This is very helpful, and has the effect of warning us against there being any single method in science. This followed by an interesting discussion of the "disunity of science," and an interesting critique of Galison's notion of the "radical particularization of science."

Chapter Four contrasts this view with two versions of the "postmodern philosophy of science:" a "pluralistic radicalization of epistemology," and an "anti-epistemological deflationist approach". The first is "responsive to the social normative conditions in which this knowledge is produced," and exemplified by the work of Longino, of whom Ginev then provides and interesting critique. Most of the chapter is devoted to a critique of the deflationist model, in particular of Rouse. Rouse's account is deflationary in that science has no "cognitive specificity." It's born of a "respect for the local context." Ginev's critique goes right to the point: Rouse "fails to understand scientific practices philosophically," for he fails to provide ways of conceiving the way theoretical knowledge persists through recontextualization. This extensive critique of Rouse is one of the clearest and best parts of the book. Rouse opposes the "legitimation project," which is the need of science to claim special knowledge of the world. Many kinds of postpositivists are involved with interpreting this project, if only to make sense of its special content. But Rouse wants to back out from the project altogether, denying that there is a special content to science. The project, Rouse says, suffers from "normative anemia." But the task, Ginev says, is not to deconstruct the claim to cognitive autonomy, as Rouse does, but to "show how scientific inquiry succeeds to create its cognitive distinctiveness within its own interpretive horizons."

In the course of the next few pages, Ginev makes a point of supreme importance – that the "democratic process of contemporary societies needs science's cognitive autonomy," lest, strangely, social life become sciencized. This broaches the issue of the role of expertise in modern society – the way that it is or is not incorporated. This issue is so important – Collins, Wynne, Jasanoff and others are wrestling with it with no clear resolution in sight — that I wish Ginev emphasized it more.

Chapter Five is a critical interpretation of the feminist critique that any epistemology reflects and reinforces the masculine point of view. It exhibits "the relevance of the doctrine of cognitive existentialism to a multi-gendered picture of science." It rejects the search for a feminist "standpoint epistemology," as well as a Rouse-like reduction of philosophy of science to cultural studies. This chapter, too, is brilliant and very revealing. Interpretation, he says, is not a "recapitulation of ready-made results of inquiry," but rather is "the ongoing fore-structure of the research process." The key insight is that discarding "representation" and "the reification of scientific knowledge", as Rouse, Rorty, and feminist cultural theorists seek to do – or "discarding the epistemological justification of an internalist view of scientific rationality" – will require disclaiming "science's cognitive autonomy." Another way to put it is that the assumption of these thinkers that the line between science and the rest of culture "can only be drawn within the framework of an essentialist conception about scientific method, language, and/or rationality" is wrong. The alternative – cognitive existentialism – involves "studying a long-term interpretative fore-structuring within a characteristic hermeneutic situation." And this is not necessarily essentialist. This is possible because "the starting point of hermeneutic philosophy is not the plurality of situated standpoints, but the reflection on the situatedness of human agents in their routine settings of actions and interactions. This is an ontological and not epistemological starting point." There is a horizon that "always already goes beyond human agents' "reality" of appropriated possibilities."

Here's the key passage from this chapter: "The degree of epistemological objectivity depends strongly on the way these presuppositions are reproduced in the configurations of interrelated practices. The more the presuppositions are involved in a domain's constitution, the weaker the epistemological objectivity is, and the more the dimension of recontextualization prevails. A characteristic hermeneutic situation specifies the research process as a process furnished by readable technologies of scientific practices that are carried out between the poles of epistemological objectivity and interpretative reflexivity" (Ginev, 2011, p. 140). This process produces cultural identity without being reducible to it.

3 Theory and Theoretical Objects

There is much more to speak about in *Tenets,* but let me stop to consider a recent critique by Arvin Vos of Ginev's approach to theory and theoretical objects.This will allow me to elaborate more thoughts on the nature of theory and theoretical concepts within an existential/hermeneutical conception of science (Vos, 2011).

Vos's principal objections to Ginev's account amount to two. A first is that Ginev places too much emphasis on practice as opposed to inquiry. Inquiry involves much more than practice, Vos suggests. Inquiry is "an ordered process," whose aim Vos seems to think is to reach "an object that identifies a class or set of things." Focusing on inquiry rather than practices, Vos says, "opens the door to a fuller exploration of the dynamic of our quest to understand." This kind of exploration "is significant because it places [theoretical objects] in a larger context, in the full sweep of our experience of knowing." Vos cites Plato as having recognized the importance of inquiry, but adds that it is not enough to say, as he suggests Plato does, that the theoretical objects discovered in inquiry were known in a prior life.

Vos's second main critique is that Ginev fails to account for how theoretical objects arise through practices. Ginev's chief problem is that to avoid essentialism he "describes theoretical objects purely in relation to practices." How, Vos asks, can practices possibly give rise to theoretical objects?

Vos concludes that Ginev is right to want to avoid essentialism but gives up too much in avoiding it; Ginev "eliminates significant features about knowing in order to avoid essentialism," especially concerning theoretical objects. Ginev "has failed to break cleanly with the fundamental assumptions about knowing on which essentialism is based." Vos therefore rejects cognitive existentialism, and offers an alternate in the spirit of Lonergan's philosophy, involving a more traditional vocabulary including words such as hypothesis, verification, and the question, "Is it so?"

Vos is correct that Ginev's account is insufficiently developed regarding these two points of critique, but fails to recognize the resources of an existential/hermeneutic conception of science for doing so.

4 Practice vs. Inquiry

Let us begin by considering practice versus inquiry. It is true that much of science, especially scientific discovery, does not merely involve *praxis* – an application of some skill or technique – but a *poiesis*, a bringing forth of a phenomenon into presence. Science is therefore much like perception, including its possibilities for error and illusion, except that scientific perception is instrumentally mediated (Crease, 1993, pp. 77–89). When the novel appears, furthermore, it usually does not do so with Cartesian clarity, but must be made to appear more clearly, so that it shows itself to us – again, through our technologically mediated practices – more as it is.

Moreover, inquiry is spurred by a particular kind of dissatisfaction within a lived situation. Such dissatisfaction takes the form, not merely of discontent or the experience that knowledge is outstanding — of a lack of understanding. There is much that I do not know that I have no desire to inquire into! Rather, the dissatisfaction of inquiry is the experience of a collision between our expectations and what we encounter, of the sense that we ought to understand something that we do not. Dissatisfaction in science arises in many ways. In each case it involves the experience of something outstanding or obscure that, we sense, can and must be brought into the open. Inquiry is a response to this experience, in which we apply what has been historically and culturally transmitted to us in our attempts to transform the dissatisfying situation. We transform the situation in making it our own, so we are more "at home" in it. In Theodore Kisiel's clever phrase, this is the "eigen-function" of research. Yet every clarification does not finalize appearances but brings new dissatisfactions and continued inquiry.

Oliver Heaviside's remark, when he took Maxwell's frustrating and torturous equations and revamped and simplified them, resulting in a huge practical and theoretical advance in the understanding of electromagnetism, that he had inherited something from Maxwell and was simply trying to take what his predecessor had done and "see it clearly" – these words could be said by any genuine researcher, including Maxwell, who took what he had inherited about electricity and magnetism and made it clear for himself (Heaviside, 1971, p. vii). We see such examples multiplied throughout the history of science. Richard Feynman created his famous diagrams for particle interactions, he said in numerous interviews, because he was confused about what he was hearing and wanted to put it in a form *he* could understand; now these diagrams has become the inherited way *we* understand.

The hermeneutical character here is that the concepts involved undergo a process of evolution as they are reinterpreted again and again in different research contexts. To understand such a continuously unfolding process, we have to pay attention to this practical process of inquiry, rather than to the beliefs or theories that are its epiphenomena or products. The world is never fully transparent, we always meet it with historically and culturally transmitted assumptions that reveal and conceal. But what appears in the acts of inquiry serves to unearth the presuppositions, expose the assumptions, that are standing in the way of the thing itself, retuning the connections between our expectations and what we encounter, and deepening our engagement (Crease, 1997).

Further complicating this process is that technologically mediated experimental performances are involved in this practical process. To pursue scientific inquiry, it is not sufficient to consult what we already have. It is not enough to read more books or talk to more people – this will not change the situation! To further practical inquiries of science, we have to stage events that show us, that give us back, more than what we put into them. What then appears in such experimental acts forces us to reinterpret our situation, recasts our resources, and reshapes our understanding. The "how" of this process includes the designing, enacting, and witnessing of events in laboratory situations (Crease, 2003, pp. 266–272). These events are not fully transparent – again, they do not deliver us objects with Cartesian clarity – nor are they permanently segmented from each other, and independent of the performances that stages them (Crease, 1998, pp. 160–168). Even theories and theoretical objects do not appear with Cartesian clarity, but need to be worked on before they are clear and consistent; here the convoluted history of quantum theory is an excellent example. When was "the quantum" introduced? Planck's version was different from Einstein's, and Bohr's, and Sommerfeld's, and so forth – and that's only to speak about the "old quantum theory." It is just as ambiguous to say when "quantum theory" was developed as it is to say (as Kuhn famously pointed out) when "oxygen was discovered."

Moreover, we grasp the meaning of such events in relation to everything else that we know; these events are implicated in our understanding of the world.

That is why what appears in the experimental acts is not merely a set of idle facts that we are free to ignore; they place claims on us. We are already related to these experimental events, even if they are enacted in special laboratory contexts. Thus the fact that these events may only take place in special laboratory contexts does not mean that they are abstract and unworldly. It's the other way around: the special laboratory contexts are what make these events part of the world, and therefore are responsible for them being of pressing concern to us. A laboratory is like a garden where special cultivation and conditions allow things to grow that may not grow "in the wild," so to speak, yet the mere existence of the things grown under such conditions makes a claim on our understanding of the "wild."

Since Plato has been invoked on the subject of inquiry, let me appeal to that example. It is, indeed, my experience in four decades of teaching philosophy that if one thinks that Plato did not get something right, one has not yet appreciated the full subtleties of Plato's account. The discussion of inquiry in question occurs in a conversation between Socrates and a youth named Meno in the eponymous dialogue. Socrates has discovered Meno to be gullible, to balk at difficult ideas, and to be impressed by theological authorities. Therefore Socrates couches his explanation of how inquiry happens inside a legend that, he says, is believed by religious sages. The point of the legend is that learning is neither like something getting handed to you by someone else, nor is it like automatically following a rule. Instead, it is an active and intensely personal process in which you collect enough experiences with what it is you are exploring, in a continually expanding knowledge base, so that when the discovery comes, it feels as if it were a feature that was already there that you had overlooked -- like you had it all along. It's like unforgetting something. *That's* why Socrates has invoked the myth about having learned things in a prior life. It's not because Socrates believes it (less still Plato); in fact, Socrates makes a point of distancing himself from the myth. It's rather Socrates' way of making the point that acquiring knowledge is not like filing things in a mental warehouse but a back-and-forth process in which we are constantly moving between parts and wholes, encounters and interpretations. In short, Plato's account is hermeneutical through and through (Crease, 2008, pp. 35–41).

5 The Origins of Theory

Vos's second objection involves a question of how theory and theoretical objects can arise through practice. How can something be discovered in practice, yet also transcend the situation involved in the inquiry?

This, too, is a topic in which hermeneutical philosophy has plenty of resources. A full elaboration is beyond the scope of this paper. Let me begin by saying that this is an area where the traditional approach of considering theories and theoretical concepts as representations leads to insoluble problems. How do theories and theoretical terms arise?

It is important to recognize at the outset that theories are not speculations or blueprints plucked from nowhere and imposed on experience. They are more like maps that tell you how to get around in places you are mostly familiar with; they connect with experience and originated in experiences. Once you discover your relatedness to a group of experiences, you want to reflect on them, and theory is the product of that reflection. It implies already having a familiarity, a knowing how to get around in the natural or experimental landscape. That implies always already having an access to that landscape.

Theories organize what we know and help us look for what we don't. Theories scale, frame, and orient. They scale by showing how to move from one level to another. They frame by giving a solid grasp of the possible and impossible in the here and now. And they orient by "indicating how we can transform the world, by suggesting what new knowledge it is important to seek, and by showing us how to go about seeking it." Theory creates a space for possible disclosure. Experimental performances actually make possible that disclosure in that open space (Crease, 2007, p. 16).

Like maps, theories and theoretical terms do not have to be complete to be true; indeed, they cannot help but be incomplete. Scientific theories and terms are usually incomplete not only when initially offered but all the way through their development; "quantum theory" is a good example, as is evolutionary theory. Scientists propose theories and theoretical terms with full awareness and expectation that these will be further elaborated by their successors to capture better what is encountered. Strange particles mutate to K particles and then to quarks and beyond – and throughout all of this it's not so much that prior theories were defective but that what they pointed to appeared differently the more we explored it. This is as true of the Higgs particle theory as it was of X rays.

It is therefore much more illuminating, I have argued, to consider theoretical concepts as something like what Heidegger called formal indicators, descriptions that are at the same time aware that they are pointings. Heidegger developed the notion of formal indicators in trying to develop a discourse that can refer to phenomena in ways that does not prejudge them but leaves open the possibility that such phenomena can appear in different ways in different contexts. Formal indicators call our attention to worldly things in ways so that others can experience them as well. The discourse is "formal" because it describes the phenomena with enough specificity that others can experience it, but only "indicates" or points at them, for these phenomena may not be experienced, or experienced confusedly, may need to be further interpreted and elaborated, and will definitely be experienced differently when our positioning (technological mediation) with respect to them changes. We recognize people and objects, Merleau-Ponty says, through their styles – but this recognition involves a more formalized process in science, one that is openended yet never leaves experience.

Formal indication involves anticipatory apprehension, and its formal aspect will surely be revised as we have closer apprehension. Theoretical terms, therefore, are not dead, decontextualized, distanced from the world. As indicators, they are instead references to what we encounter. "Might not scientists, in practice, hear the language of their discourse not as ground-plans of nature, not as "cognitive essences" that are operationalized, but as indications allowing them to pick out phenomena which become bodily present in instrumentally mediated ways with possibilities of appearing otherwise given other mediations? … What if, in short, the way scientists use conceptual language does not aim to grasp a decontextualized object, but rather [involves] indication and fulfillment?" (Crease, 2009, p. 228)

Furthermore, one must beware of the dual semantics of theoretical terms, for the same term can name one thing in a theory (electron) and another in the laboratory (the worldly, measured electron). This is like the dual semantics of musical terms, where a 'C' can refer to a (theoretical) note in a score, as well as a (performed) note heard in a concert hall. In the scientific context, there is a back-and-forth between those two senses, for the electrons that appear in the laboratory may lead us to revise the electrons in the theory, which will lead to new experimental stagings and new appearances of electrons in the laboratory, and hence new electrons in the theory, and so forth. The point in a hermeneutical perspective is that the worldly appearances have priority; in fact, the theories (even the musical scores!) were formed to produce and formally indicate the worldly appearances.

How do these theories and theoretical terms, which help us to navigate our laboratory experiences, arise? They arise out of our practices as ways to express our grip on the world. Let us take, again, Plato's example. The key moment – the moment when it looks as though Socrates is plucking something from beyond the slave's experience and pasting it on what they are talking about – occurs when Socrates adds the diagonal to the diagram that until then has consisted all of squares. But the diagonal is not entirely new! The slave has seen diagonals in floor mosaics and wall designs, and is merely being reminded of what one is. This diagonal casts the problem in a larger, richer context that allows the slave to reconstruct the problem. It is a new term, but one familiar to the slave from his own experiences. If he weren't, the term would be useless.

This invocation of a piece of familiar experience in an inquiry resembles the way theory develops by borrowings from familiar practices, an extension of the old Vichian principle that human beings understand the unfamiliar by the familiar. This is the spirit of Jeremy Bernstein's remark that "it is probably no exaggeration to say that all of theoretical physics proceeds by analogy" (Bernstein, 1968, p. vii). It is also related to Einstein's observation that the relation of theory to data is "not analogous to that of soup to beef but rather of wardrobe number to overcoat" (Einstein, 1950, p. 64). There's no magic involved in the application of numbers produced in experimentation to phenomena; they are a function of our practices in how we've built and arranged the equipment.

Finally, the notion of formal indication does justice to the temporal character of the scientific process, to its relation to its own past – however much it might seem like "beautiful ruins" — and to its evolving future.

The real is what we encounter in the present — what we can confidently and even unavoidably reach, what appears in our horizon inevitably, and what we cannot turn away from. Yet the present seems also a little unclear, somewhat discordant, not fully grasped, with hints of another, deeper order just over the horizon. This discordance is why we inquire, and what makes newly achieved discoveries seem, strangely, to be both discovered and invented. And when these discoveries arise, the greater unity thus achieved promises to turn our present into ruins (Crease, 2009, p. 227).

That is why I have argued that one requirement for an adequate philosophy of science is "covariant realism," or the constraint that the real cannot be the function of a single theory, or one historical moment, which is extended only charitably or metaphorically to other moments in science history. The real must not be the function of representation but of our experiences (*ibid.*).

In an existential/hermeneutic perspective, theory and theoretical terms do not so much answer the question "Is it so?" as point to where and how such a question is to be answered – in the density and dynamism of our experiences. Such a perspective reveals the stunning groundlessness of theories and theoretical terms, which we adhere to, or not, because of their ability to create a space for continued disclosure of the technologically mediated world we encounter.

WORKS CITED

Bernstein, Jeremy. (1968), *Elementary Particles and Their Currents*, San Francisco: W. H. Freeman.

Crease, Robert, P. (1993) Crease, Robert, P; *The Play of Nature: Experimentation as Performance*, Bloomington: Indiana University Press.

————. (1997) *Hermeneutics and the Natural Sciences*, ed. R. Crease, Kluwer.

————. (1998). "What is an Artifact?" *Philosophy Today*, SPEP Supplement.

————. (2003) "Inquiry and Performance: Analogies and Identities between the Arts and the Sciences," *Interdisciplinary Science Reviews* 28:4

————. (2007) "Just a theory," *Physics World* June.

————. (2008) *The Great Equations*, New York: Norton.

————. (2009).Covariant Realism," *Human Affairs* 19, 223–232.

Einstein, Albert. (1950) *Out of My Later Years*. New York: Philosophical Library.

Ginev, Dimitri. (2011) *The Tenets of Cognitive Existentialism*, Athens: Ohio University Press.

Heaviside, Oliver. (1971) *Electromagnetic Theory*, vol. 1 (New York: Chelsea).

Vos, Arvin. (2011) "On the Status of Theoretical Objects in Science according to Cognitive Existentialism," *Balkan Journal of Philosophy* 2.

Three

PRACTICE THEORIES: SCIENTIFIC CHARMS, DIVINE SPELLS

Evaldas Juozelis

The mission statement of *Society for philosophy of science in practice* nicely acquaints to the scope and capacity of practice approach regarding philosophical problems of contemporary scientific edifice:

> Philosophy of science has traditionally focused on the relation between scientific theories and the world, at the risk of disregarding scientific practice. In social studies of science and technology, the predominant tendency has been to pay attention to scientific practice and its relation to theories, sometimes wilfully disregarding the world except as a product of social construction. Both approaches have their merits, but they each offer only a limited view, neglecting some essential aspects of science. We advocate a philosophy of scientific practice, based on an analytic framework that takes into consideration theory, practice and the world simultaneously (Society for Philosophy of Science in Practice, 2014).

Two perspectives I ought to stress here and to converge them into practice theories forthwith: first, the notice of necessary restrictions inherent in "traditional" and "social" philosophies of science, and second, the promise not to simply elaborate common themes characteristic to previous endeavors, but to amend them at a substantial level while invoking "the world". Quite intriguingly *the world* appears to be the most lacking part of practice theories' philosophical associates, despite the obvious – it must be, one way or another, a point of departure and ultimate goal of any scientifically, i.e. seriously inclined philosophical talk. If we have too much of *theory* or too much of *practice* – eventually, our "analytic framework" does not work properly. This is why the Society addresses some general principles for research programs that contrary to their predecessors engage the world simultaneously merging knowledge with its purpose, conceptual tools with laboratory machinery, individual qualities with political fibre, as well as cultural strands with biophysical environment. If any kind of knowledge is not separable from its aims and concreteness to be fully grasped, an integral practice theory, no matter under a label of naturalism, pragmatism or cultural studies should be based on practice as implicitly constructive ontological source.

I hold Theodore Schatzki's *Practice theory* one of the best introductory texts concerning subject matter of practice approach. Having in mind the diversity of so-called practice accounts he provides convincing assumptions determining the apparent tension between traditional standpoint and practice attitude. For Schatzki, practice theories (even though practice thinkers are naturally sceptical about *theory* as an instrument capable of precisely depicting or explaining practice as such) present human activity as intimately materially engaging the world in a way that inextricably joints understanding and embodiment. Ongoing experiencing of immediate environment and its performativity thereof constitutes bodily activity framed within practical context. At this phase, practice approach takes up a broadly materialist bearing to inculcate nonhuman agency into practice-oriented structuring of the world. Thus, practice thinker (or thought practitioner) finds himself in (or rather is captured by) a process of reciprocally mediated realm of, say, objective/subjective, material/spiritual, factual/speculative, natural/normative, etc. However, somehow paradoxically (and imperatively) practice thinker is susceptible of actively issuing those distinctions and is amply aware of it.

To be more exact, Schatzki describes practice theory as a platform firmly rooted in "material configurations" and "nonhuman entities". The point is that practice theory so reconfigured may take an initiative to revise traditionally sound distinctions on the grounds of practices that are neither identifiable with nor reducible to human acts, social sphere, theoretically designed compositions of the natural, or an arrangement of them all. If "nonhumans do not just mediate, but themselves propagate practices" so that practices "comprise human and nonhuman activities", practice invariance, in other words, casts doubt on any practice analysis that is putatively substantiated through articulation of essential differences between human and nonhuman agency. Accordingly, the primacy of practice is supposed to surmount distinctions of mind/matter, body/self, nature/culture, etc. employing performativist, nonrepresentationalist strategy. Therefore reason, language, actions, individuals, institutions, life worlds and so on are themselves practice phenomena that have only derivative weight, and are subsumed within practice field by practical understanding (Schatzki, 2001, pp. 10–23).

In this light, I single out practice theories' radical posthumanist party, as it virtually stands alone even among fellow practice co-thinkers, who do not necessarily adopt performative idiom and hardly are nonrepresentationalists or antiessentialists. Posthumanists (I consider Andrew Pickering and Joseph Rouse the foremost deputies among them) typically target at scientific realism, antirealism, constructivism, to a certain extent hermeneuticism and phenomenology, and seemingly push practice approach agenda to its extreme logical end. The assumed shortcomings of philosophies of science involved in Cartesian oppositions are settled here by disentangling practice from its imaginary subordinate bondage to humans. Furthermore, posthumanism does not turn into nonhumanism or anti-humanism either, for engaging the world practically would be exactly the way out of human-nonhuman confrontation.

So two ultimate questions emerge after all: what kind of "metaphysics" of the world posthumanism promulgates, and what role it provides for a knowing subject – bodily thought practitioner?

As for the first issue, one must at least count pragmatism and/or Heideggerian ontology in, as for the second issue similarly, though extrapolating "metaphysics" into Merleau-Pontian phenomenology and/or Foucauldian genealogy, and then interpolating it back. Agenda of this kind is fairly noticeable in Rouse's philosophy of scientific practices. If one is decisively committed to overcome binary positionings of theory-world (disregarding practice) or theory-practice (disregarding the world), the best option is not to get involved into the same strategy. Instead of working out how to rename speculative problems or correct tactical errors of philosophical counterparts (say, in bridging epistemic gaps left of binary divide in "a new way"), posthumanism takes full risk to change the old strategy leaving behind its preconditions, and herewith rather safely embraces its merits.

Even those who insist that language and culture must always "interact" with their material surroundings still start from the pervasive and problematic conception of distinct components of the world that interact rather than a complex field of material-discursive practice. /.../ Perhaps we can begin to learn from the still undead debates over social constructivism how to drive a stake through the heart of such assumptions once and for all (Rouse, 2002, pp. 77–78).

The lesson to be learned concerns not only language or culture usually describable in interactionist (therefore representational and constructivist) terms, but the knower as well. Whenever he or she is caught in a trap of representational epistemology, the knowing subject loses grip of "how the world responds to human intervention and thereby orders human activity" (Schatzki, 2001, p. 16). Hence Pickering talks of scientific practice as cultural extension that "includes the traditional one of understanding how new knowledge is produced in science, but goes beyond it in its interest in the transformation of the material and social dimensions of science" (Pickering, 1995, p. 3). If representational idiom plunges science into an activity of scientists as "disembodied intellects" that allegedly produce knowledge of "how the world really is", an alternative starting point is "the idea that the world is filled not, in the first instance, with facts and observations, but with agency" (*ibid.*, p. 6). Pickering states further that in a performative idiom "science is regarded a field of powers, capacities, and performances, situated in machinic captures of material agency". That is, human agents construct machines, which "capture, seduce, download, recruit, enroll, or materialize that agency, taming and domesticating it, putting it at our service". However, I would add that nonhuman agency, or a field where scientific practitioners "maneuver", captures, seduces, materializes and domesticates human agency itself. Indeed, if the world according to Pickering "is continually doing things" that bear upon us "as forces upon material beings", we obtain sufficiently unequivocal metaphysical picture. Contrary to representationalist account, reducing science (or scientific practice) to either human activity or material settings, Pickering's temporal emergentism dissolves them into "constitutive intertwining". Yet human agency and material agency "are not continuously deformable into one another", but rather "are intimately connected with one another, reciprocally and emergently defining and sustaining each other" (*ibid.*, p. 17). Still contrary to phenomenological narrative, the intentional character of scientific practices seems to be of no primary importance to him, because it evidently disperses along temporally emerging situativeness and performativity of practice dominion. (*ibid.*, p. 19–20, 195). Sequentially, inasmuch as "practice consists in the reciprocal tuning of human and material agency", whereas tuning "can itself reconfigure human intentions", Pickering concludingly offers a metaphor of a "dance of agency".

The dance of agency, seen asymmetrically from the human end, thus takes the form of a dialectic of resistance and accommodation, where resistance denotes the failure to achieve an intended capture of agency in practice, and accommodation an active human strategy of response to resistance, which can include revisions to goals and intentions as well as to the material form of the machine in question and to the human frame of gestures and social relations that surround it (*ibid.*, p. 22).

So nothing too important is left to say here of being bodily knower, except that the desirable symmetry of human-nonhuman is available within dialectical "mangling" of resistance and accommodation. Despite indispensable simplification of Pickering's attitude just have made, I hope I haven't distorted the substantial principle of posthumanism I reckon particularly relevant to my consideration on practice approach and its inviolable trinity – practice, theory and the world. Practical understanding – oxymoron as it is – is always a possibility, which is never explained away or is efficiently stabilized precisely for the reasons implicit in a "genealogy" of practice comportment. "Real-time practice" constantly cheats upon theoretical control "offensively" evading, as Pickering puts it, "deeply ingrained patterns of thought".

The world of the mangle lacks the comforting causality of traditional physics or engineering, or of sociology for that matter, with its traditional repertoire of enduring causes (interests) and constraints (*ibid.*, p. 24).

That is, if in a posthumanist space "the human actors are still there but now inextricably entangled with the nonhuman, no longer at the center of the action and calling the shots" (*ibid.*, p. 26), typical interpretation of practical understanding as nevertheless embodied in individuals as social agents is deprived of its rigor. Moreover, if humans and nonhumans are ontologically equal in respect of practice field, it arguably connotes ontological intertwining of what is usually expressed as "the natural" and "the normative" (Schatzki, 2001, p. 18). Stringent representationist division of the normative and the natural realms still prevalent even in practice theories is precisely the target of Rousean philosophy of scientific practices, but also the master-key to his own practice approach envisioned to "drive a stake through the heart" of representationalism. The recipe is plain and simple.

Once the normative and the natural or causal dimensions of the world are thus conceived incommensurably as self-enclosed domains, however, they can never effectively be brought back together. We must understand causality as always already normative, and normativity as always already causally efficacious (Rouse, 1999).

The realization of it is a very complicated story though. As one could expect from the posthumanist point of view the problem is that the sovereign "humanity" unilaterally autonomizes and institutionalizes various human activities simulating techno-scientific solutions, but ignores the fact that science does not represent humanity "as such" (because, in short, there is no science "as such"). Sets of practices that constitute sciences, although conventionally considered as separate disciplines, "materialize" only engaging local milieu, so the soil is not conducive to preserve any "deeply ingrained patterns of thought" or universal foundations, rather the opposite. On the one hand, antiessentialist nonrepresentationalist concept of science relieves of some mostly functionally defined general obligations usually attributed to science; on the other hand, it demands additional arguments of how performative narrative of posthumanist practice approach outmaneuvers old-style philosophy and substantially amends it. In both cases, the issue is a manufacturing of scientific activity, its circulation and use for certain purposes by certain agents: scientific activity performs a number of functions and inevitably *explains* and *is explained* in one way or another. But Rouse perhaps would argue that there are considerable differences, which arise from the spatiotemporal structure of posthumanist narrative: scientific practice theory focuses on prospectivity (instead of retrospectivity), interpretativity (instead of explanation), situativity (instead of universality), and discursivity (instead of stability). Functional representationalism is therefore "humanist": empiricist or rationalistic, antirealist or realistic (in sum – legitimistic); practice approach is "pragmatist", "naturalistic" and "posthumanist", and requires some sort of mixture of cognitive and biological, normative and causal, cultural and natural factors. The fusion obtained emanates another type of individuals, who, as Schatzki notes, "instead of effecting and sustaining norms, meaning, and language out of their own resources, are integrated (to varying degrees) into the ways of proceeding that characterize extant practices, where these matters are conserved and novelty and transformation take their start" (Schatzki, 2001, p. 18).

Bipolar naturalized narrative implies both materiality of practices and causality of discursive articulation as corporeality of conceptual normativity, hence the sovereignty of "human factors" (will, intentions, reason, etc.) is held excessive and inadequate in respect of natural environment. Altogether, modernist representationalism operates within epistemic and sociological resources and aprioristically identifies the homogenization of methodological principles, pragmatic technological goals and social consensuses with scientific activity; in Rouse's practice approach exactly the general scientific assumptions, characterization of activities' scientificity, identity of scientific communities are problematic locale absent of justification and in turn encourage to stress the heterogeneity of practice medium that requires ongoing efforts to reconstruct the meaning and significance of practices. Accordingly, in a broader context of the narrative naturalization temporal dimension translates biological, social, and cultural distinctions into conventional categories; more narrowly, questionable are considered any efforts not only to separate scientific activities from social and cultural environment, but to interpret practice as self-justifying socially organized acts, rational decisions or individual capacities ("humanity", "methods", or "experience"). If you follow antiessentialist guidelines, then you do not have major problems to transform into science any form of knowledge or practice – all you need is to properly place it in a "discourse" and reproduce it *in situ*, by no means *a priori*. I trust the practice theory of Rouse skillfully outmaneuvers "traditionalism" so that instead of constructing another narrative of progress it implies the progress of narrative – the change of the story sensitive to practice itself.

Now what is important in the present context and what is actually the most intriguing part of the practice narrative personally to me, exposes the line of argument of hermeneutic phenomenology, to be precise, Dimitri Ginev's cognitive existentialism. The essence of cognitive existentialism as I see it consists of a set of critical claims whose purpose is analogous to those elaborated within certain (especially posthumanist) practice theories designed to assess critically and supplement the established philosophy of science. The difference is that Ginev's cognitive existentialism focuses not so much on the traditional philosophy as on practice turn in philosophy, and in fact does to the practice theory what practice theory has been initially committed to do to traditionalists. Consequentially Ginev steps into the Practice-Theory-World magical triplex resonating to the most part of the practice approach labourers' campaign, yet waving the banners of hermeneutics and phenomenology. As if echoing Rouse's contemplation of the world as "complex field of material-discursive practice", Ginev declares:

> [T]he theory of the discursive practices taking place in scientific research forms an independent context of studying science. It is this "context of constitution" in which one thematizes the hermeneutic fore-structure of scientific research. /…/ Starting assumption of these studies is the ontological priority of the "practical horizon" of being involved in a research process over the theoretical attitude; the studies are neither empirical nor normative, but interpreting the proto-normativity of hermeneutic circles that are constitutive for scientific research; the studies are scrutinizing scientific research in its "effective-historical" interpretative openness (Ginev, 1999).

What is this all about I will try to explicate as briefly as I can, naturally taking full responsibility for the very likely misrepresentation and underestimation of cognitive existentialism. My own starting assumption is that just as any representative of the ideas of *Society for philosophy of science in practice* believes in his or her knowhow to point out and revise the deficiencies of traditional philosophies of science, so is Ginev the believer in cognitive existentialism as the unerring movement forward highlighting and solving the problems of both traditionalism and the practice approach. Nevertheless, even greater difference lies not in the scope or ambitions of the believers' attitudes, but the preferred styles of philosophizing. Practice thinker might say as I previously cited: traditionalism has its merits, but offers a limited view because neglects some essential aspects of science, therefore I advocate…; Ginev says: "Rouse's cultural studies of science do not reconstruct the problematics of philosophy of science in any critical way. They simply trivialize this problematics"; or: "If by claiming (for good reasons) that there is no intrinsic 'epistemic status' of scientific activities, one goes so far as to rule out the problem of the constitution of cognitive structures (and research domains), then one inevitably trivializes scientific activities" (Ginev, 2006). Why so?

> [B]oth the schools of social studies of the natural sciences and the programs which Andrew Pickering unifies under the heading "science as practice and culture" did not succeed in finding a reliable (ontological, epistemological, and methodological) solution to the problem of relativism. Neither "naturalizing the relativist mode of investigation" nor the rejection of a "metanarrative" combined with the search for "new literary forms" promise acceptable solutions. From the viewpoint of hermeneutic phenomenology, the formulation and solution of the problem of relativism requires a conception of the situated transcendence of human practices (*Ibid.* 2006).

"The context of constitution", "hermeneutic fore-structure", "proto-normativity", "interpretative openness", "situated transcendence" are crucial but not the sole tools in cognitive existentialism's toolkit devised to show how really deep human understanding in the light of hermeneutic phenomenology is. Cultural studies of science, naturalism, relativism, constructivism, and so on are philosophically shallow. Self-esteemed hermeneutical – therefore true – philosopher could never utter something like underthinker Pickering in his confession that

> [I]n the analysis of real-time practice, in certain respects at least, none [explanations] can be given. I can do nothing about this, but it is best to be clear on this point from the start (Pickering, 1995).

Surely, it is an unacceptable solution for a transparent mind capable to dissociate himself even from the "practical horizon" thanks to the situated transcendence veraciously allotting practical and theoretical attitudes on ontological grounds. First, to achieve the required solution Ginev scrutinizes the "dominant account" of "the essentialist understanding of science". The essentialist understanding in the forms of naturalized epistemology or normative epistemology acquire doctrinary types in terms of scientific method, truth of science, objectivity of scientific knowledge, rationality of science, and privileged status of science in modern societies respectively. Set apart, these (thus non-radical) types of essentialism are vulnerable to and virtually compatible with their (non-radical) antiessentialist surrogates: cognitive relativism, (methodological, epistemological, or social) constructivism, and various sorts of Finean natural ontological attitude, Rortian neopragmatism, or Feyerabendian political philosophy of science (Ginev, 2006, pp. 75–83). Implicitly, cognitive existentialism is not another version of essentialist-antiessentialist blunder; it is a radical antiessentialist antagonist whose philosophical resources pour out of entirely different ground, called existence.

> By "existentialism" I mean, in the present context, a "radically antiessentialist" viewpoint whose main characteristic is the search for the "existential genesis" of every supposed "essence" in human life. /…/ To define science's cognitive specificity from this viewpoint amounts to revealing scientific research as a mode of being-in-the-world based upon a specific "existential project" (i.e., a distinct way of interpretative realization of possibilities already projected [*Ibid.* p. 85]).

Then, Ginev couples existential conception of science with hermeneutic phenomenology, i.e. pre-epistemological constitution of meaning that has ontological as well as transcendental dimensions. Since this constitution occurs "always already" within the world, "the world 'transcends' all activities of every particular epistemic subject" turning the constitution of meaning into a "trans-subjective process". The trans-subjective process as a "primordial trans-subjectivity of existence" is in line with the "interrelatedness of practices", which is associated with "the world as a horizon". Whereas the interrelated practices exemplify "each form of human existence" as a "particular kind of subjectivity", the constitution of meaning as an articulation ("understanding") of any kind of subjectivity ("as expressed in a potentially infinite variety of meaningful contexts") is the underlying activity of the swiftly revolving hermeneutic world, since "'understanding' connotes the way one understands oneself in accordance with the possibilities of dealing with those entities which one has at hand when one is involved in particular contexts of interrelated practices". It also means that "one constantly interprets one's understanding (one's horizon of possibilities as projected by the interrelatedness of practices in which one is involved)", and, most importantly, that "the interpretative appropriation of possibilities brings to the fore an ontological dimension" (Ginev, 2009, pp. 365–366).

One might concern right away that Ginev's hermeneutic universe fashioned in this style contains not only Heideggerian "existential conception of science", but also a kind of Nietzschean interpretativism. For "there is *no 'essence'* behind or beyond the interrelatedness of practices"; "the being of human existence is revealed by means of the *interpretative understanding* that articulates the meaning in the world"; "there is *no being* but being-in-the-world", "the interrelatedness of practices reveals by necessity *not a static being* but a potentiality-for-being" (*Ibid.*, [itallics by Evaldas Juozelis]). Yet, according to Ginev, "the situatedness of scientific knowledge does not require either deflationary [Rousean] accounts or a Nietzschean naturalism" (Ginev, 2005, p. 193). As one could foresee, Rouse's "quasi-Nietzschean account" (formulated in "quasi-Heideggerian" terminology) fatally lacks the depth of hermeneutic ontology, for it deliberately ignores the "ontico-ontological difference". Perhaps analogous deficiency should be attributed to Nietzsche too.

> The world in which human beings are 'always already thrown' is the horizon of human beings' discursive practices. It is no longer the natural world (that can be presumed in whatever kind of naturalism [*Ibid.* p. 206]).

Instead, in the existential conception of science, "the transcendence of the world (a transcendence which is grounded in the 'ecstatical unity of temporality') makes it possible to sketch out the way of objectifying". Hermeneutic transcendentalism is very subtle indeed: ontologically, any isolated transcendental position must be rejected, since there are no objects beyond the "horizonal temporality" of human existence. In contrast to all possible depictions of the world by any *philosophical naturalism* (what is, by the way, *contradictio in adjecto* for hermeneutic consciousness), the pro-Heideggerian world "is the complexity of organized equipment and practices in which human beings are involved", or put differently, it "does not have an essence behind the organized equipment and practices". Forasmuch as "the transcendentalism of ontico-ontological difference is indispensable for the constitutional analysis of scientific research" regardless of its "ecstatical" nature (because "the ontological difference (qua the transcendental dimension of this analysis) is always to be taken into account when one is trying to develop a hermeneutic phenomenology of the natural sciences"), one also must note that "in carrying out this analysis the transcendental dimension has to be minimized as much as possible" (*Ibid.*, p. 369, 372).

Here finally, hermeneutic phenomenology of science opens up a path to cognitive existentialism – the ultimate stage of post-metaphysical hunting for the natural sciences' temporarily lost identity. Leaning upon the three "achievements" of hermeneutic phenomenology, namely the "de-transcendentalization" of metaphysical ontico-ontological difference, the rejection of representationalism in the analysis of scientific knowledge, and the successful defence of the natural sciences' autonomy and cognitive specificity extracted from "the modes of scientific communities' being in the worlds of scientific research", cognitive existentialism proudly informs that "only within the changing configurations of scientific practices" the sciences acquire the possibilities to do research, and that those possibilities are "contextual" and "situational", and "do not exist *per se*". Moreover, "the horizon of possibilities is always already transcendent as regards the actual situations of scientific research", but the science practitioner shouldn't worry about the transcendence, for "neither the horizon nor the particular configurations in which the possibilities become appropriated has/have a temporal priority"; yes, "they are mutually dependent", but the practitioner shouldn't worry about that either, since "there is no causal relationship but a hermeneutic circle of co-dependence" (*Ibid.*, pp. 377–378). In sum:

> [T]he unity of a given mode of being in a world of everyday practices, in which the possibility of the meaningful articulation of that world rests, is a unity that consists in the interpretative appropriation of the projected horizon of self-understanding (*Ibid.*, p. 381).

Hereby the radical antiessentialism of cognitive existentialism seems to put everything in its proper place. If one still cannot comprehend what kind of the world or the knower cognitive existentialism talks about, or how in this rapidly revolving co-dependant system I accomplish self-understanding, it is because of the delicate "structure of self-referentiality": the world as well as the knower (me) is what I understand in terms of the possibilities that actually are not possibilities but eventual configurations of trans-subjective practices to whom I always already belong. Naturally, it is only my interpretation of Ginev's statements "one understands oneself with respect to the possibilities one finds always already projected in one's practices", and "self-referentiality taking place in the nexus of understanding and interpretation is a trans-subjective phenomenon" (*Ibid.*, p. 384). Let me briefly refer back to Schatzki's article sketching disagreements between "humanist" and "posthumanist" practice thinkers: humanist theorists follow their rebellious doubles in acknowledging human agency's close ties to "'external' arrangements of humans and nonhumans"; however, they remain convicted that "these facts neither replace human agency with, nor fragment it into, the actions of these systems or arrangements" (Schatzki, 2001, p. 20). Now I feel perplexed whether I should assign Ginev to humanist or posthumanist faction. Philosophically, he belongs to none of them. Specifically ("in practice"), Ginev propagates the practice approach in developing an account of practices, therefore I do not feel uncomfortable to inquire what sort of human-nonhuman mix his account of practices implies. Usually phenomenology shows no associations with any kind of "posthumanism". However, cognitive existentialism is not (as one can easily notice) the usual phenomenology and I again want to draw attention to the style of Ginev's hermeneutical-phenomenological thinking.

Ginev frankly admits Heidegger's significant influence on his own phenomenology of scientific practices and abundantly exercises it. Often Heideggerian spirit dwells so extensively I must avow I find it troubling to discern where Ginev speaks his own mind. Yet I trust they both share not only the selfless commitment to apply "transcendental hermeneutics" to all genres of philosophy or cultural practices which have a "structure of self-referentiality", but the belief in the inevitably ontological character of the application process (Ginev, 2009, p. 383). It *should be* that way if one is capable to *understand* and simultaneously to *act*. From this point of view, any essentialist ontology (i.e. any kind of pragmatism, naturalism, realism, constructivism and so forth, despite their formal "antiessentialism") is "'derivable' from the existential ontology of the meaningful world that always already transcends each contextual configuration of human practices" (*Ibid*., p. 367). This is not a matter of intellectual preferences, favourable method or cultural setting, for "there is no human activity which is not hermeneutically fore-structured" or "predicated on a dialogical constitution of meaning" (Ginev, 1999, p. 149). On the other hand, nothing inadmissible is being said here, since the universalization of hermeneutics brings no universal (historical, evolutionary or whatever) restrictions, invariants or regulations – nothing of the essence. A careful reader might ask then what is *de facto* use of hermeneutic phenomenology beside the *de jure* acknowledgment of transcendental depths of human understanding. In other words, how it actually outrivals aforementioned practice theories wrestling with the question: "How does the richly meaningful, normative field of human activity ever successfully connect to 'nature', the seemingly semantically inert inexorability of physical regularities or causal processes?" (Rouse, 1999). Indeed, one of the focal points of it is simply to demonstrate how "understanding" is able to draw the line beyond which "the natural" mind cannot operate. What "exists" beyond the line seems to be an empty question even for hermeneutics, nevertheless it proudly announces that, say, coherent naturalism even cannot presume it. But "ontologically" Ginev's Heideggerianism risks a great deal. To be sure, naturalistically inclined philosophical thought rarely escapes a certain amount of honest naiveté, what brings it closer to a "serious" scientific discourse along with its simplicity, apparent shortcomings and persistent though inarticulable desire to stay on some solid ground. (It would be my resume of posthumanism if in one sentence). Heideggerianism in my eyes is much less attractive, arrogant, complicated, fatalistic, and intolerant.

To explain my discomfort, I refer to Algis Mickunas, a phenomenologist "of the first sort". He reminds us that by Eighteenth century reality was defined not in terms of its being but in accordance with "the conditions for the possibility of being". Astounding as it is, wonders Mickunas, now the majority favourably speaks of Heidegger's "breakthrough" in "philosophy". It is Heidegger who "discovered" that "human being in the world is primarily concerned with the 'horizon of possibilities' of being, wherein every thing in the environment belongs in a system of connected meanings wherein things in hand lose their thing character and point to other things in their usage as implements". This means, Mickunas clarifies, that *a priori* we are in a modern world of *ta pragma*, where every thing (from hammers to wild berries…) "has a meaning insofar as it belongs to our world of circumspection", and "the possibilities that are open to all are those of *Dasein* who is always and already projected ahead of himself, and hence before he can encounter anything in the environment in its own right, as 'this there,' as *to de ti,* he already draws it into the sphere of his horizon" (Mickunas, 2012, p. 65). For those not familiar with Murphy's laws it is worth making acquaintance here with Baruch's observation: If all you have is a hammer, every problem looks like a nail. Nonetheless, the upshot of the Heideggerian narrative is enlightening.

> [A]t the end [Heidegger] reverted back to Mid-Eastern cult of personality, proclaiming that "Only God can save us now," reverting back to modern neo-scholasticism. While demanding the modern surrender of "humanism," Heidegger demanded the subjection of man to the decisions of a paternal divinity. Is he not reverting to the unphilosophical tradition of the metaphysics of the will against which he attempted to exert all of his efforts? After all, if empty temporal possibilities – even those of Being – are ahead of and determine the very understanding of the things of the environment, then we are granting priority to the primacy of the will (*Ibid.*).

Similar rhetorical questions I would readdress to Ginev: do the shortcomings of Heidegger's conception of science he identified and all the efforts he put to overcome them save cognitive existentialism from superfluous omniscience; does human intentionality thrown into trans-subjective practice-field still carry the easiness of phenomenological perceptuality; is situated transcendence not yet another manifestation of the primacy of the will?

Symptoms of the longing for paternal divinity Mickunas envisioned in Heidegger's thinking beg for an alternative to existential ontology as the universalization of instrumental rationality originally cloaked in the discourse of solicitude for human existence. I am not going to discuss what sort of salvation Mickunas himself proposes. However, I invite to consider an approach implied in his harsh opposition to Heideggerianism that has direct ties with posthumanism of Pickering and Rouse. If succinctly, a Nietzschean alternative to the structuring of "human existence" in terms of "interrelatedness of practices" and "particular kind of subjectivity", and to the hammernailing Heideggerian world in terms of "complexity of organized equipment", would be the cosmos, where "cult of personality" with all the divine "shadows" is displaced. Nietzschean cosmology turns us back to a kind of naturalism that doesn't take itself too seriously, i.e. gives way to an "aesthetic" discourse that sustains materiality, historicity, sensitivity to "the world" and to our own organic nature, yet is intrinsically ambiguous, paradoxical, rejects fundamental separation of natural world and humanity, promotes "better" interpretations and still is susceptible to its own falsity. In the context of Nietzschean naturalist program of the "naturalization of humanity" and the "de-deification of nature" posthumanist perspective of practice theories might emerge as a resilient platform, prepared, as Pickering hopes, to face up "a sense of brute chance" integral to practice, and capable (I hope so) of dancingly driving another stake, but this time through Heideggerian divine complex (Cox, 1999).

WORKS CITED

Cox, Christopher (1999). *Nietzsche: Naturalism and Interpretation.* Berkeley: University of California Press.

Ginev, Dimitri.(1999) *On the Hermeneutic Fore-structure of Scientific Research* Continental Philosophy Review, No. 32.

————. (2005) *Against the Politics of Postmodern Philosophy of Science* International Studies in the Philosophy of Science, Vol. 19, No. 2.

————. (2006) *The Context of Constitution. Beyond the Edge of Epistemological Justification* Boston Studies in the Philosophy of Science, Vol. 247.

————. (2009) *From Existential Conception of Science to Hermeneutic Phenomenology of Scientific Research* Journal of Philosophical Research, Vol. 34.

Pickering, Andrew. (1995) *The Mangle of Practice: Time, Agency and Science.* Chicago: University of Chicago Press.

Rouse, Josepht. (1999)*Understanding Scientific Practices: Cultural Studies of Science as a Philosophical Program*, 1999, p. 2. https://wesfiles.wesleyan.edu/home/jrouse/Understanding%20Scientific%20Practices%20(SSR).pdf

————. (2012) *Vampires: Social Constructivism, Realism, and Other Philosophical Undead* History and Theory, No. 41, 2002.

Schatzki, Ted. (2001) *Practice Theory* In: *The Practice Turn in Contemporary Theory.* Ed. T. Schatzki, K. Knorr-Cetina, and E. von Savigny. London, N.Y.: Routledge, pp. 10–23.

Society for Philosophy of Science in Practice. (2014). Mission Statement, http://www.philosophy-science-practice.org/en/mission-statement/

Mickunas, Algis (2012). *The Divine Complex and Free Thinking.* NY: Hampton Press, 2012.

Four

REJECTING CARTESIAN CUTS: CHOOSING SELLARS OVER GINEV'S HEIDEGGER

Fabio Gironi

My review of Dimitri Ginev's *The Tenets of Cognitive Existentialism*, ended with this paragraph:

> [i]n the final analysis...Ginev's book is a skilfully crafted—if at times dense—account of a hermeneutic-phenomenological philosophy of science (far more nuanced than my brief account could describe), worthy of a placement along the spectrum of possible approaches to the problems raised by the practices of the natural sciences. Both its main strength and weakness have a common source: the adherence and commitment to the post-Heideggerian project of hermeneutic phenomenology. Those already persuaded by its methodological viability as a philosophical approach tout court will find in Ginev's book a strong development—and improvement—of it as specifically applied to the scientific sphere of human concernful dealings; those rejecting (for various reasons) its emphasis on the inescapable thrownness of practically engaged human beings into an always already pre-constituted or fore-structured "world" (and perhaps more keen to pursue a robustly realist philosophy of science) will find in [*The Tenets of Cognitive Existentialism*], or so I contend, too little to be persuaded to change their mind. (Gironi, 2012a, p. 464)

My aim in what follows is to elaborate this claim, to explain the reasons for which I reject Ginev's theoretical foundations, and to propose an alternative picture. My focus, therefore, will be on Ginev's Heideggerian legacy. Ginev's disagreements with Heidegger *vis-à-vis* the hermeneutic circumspection proper of scientific everydayness notwithstanding, his kind of constitutional analysis, informed by late twentieth-century hermeneutic phenomenology of science, shares the basic rejection of representationalist epistemology and scientific objectivism that motivated Heidegger's own revolutionary approach to long-standing philosophical debates. The internal coherence and robustness of Ginev's philosophy of science, as I observed in my review, are evident in its ability to account consistently for the complex ensemble of theoretical and experimental activities that make up the natural sciences. It is, however, the founding phenomenological-hermeneutic axioms which I find objectionable, refractory as they are to the robust realist distinction between epistemological conditions and the epistemically unconditioned real which I take to be not merely the correct philosophical (or meta-scientific) outlook on the universe we happen to be part of, but also a non-negotiable precondition for the socially and cognitively emancipatory interpretation and application of scientific research.

To describe the playing field of the philosophy of science as framed by and limited to the two opposite poles of "standard" scientific realism and hermeneutic realism (as for example is done by Sankey and Ginev map-drawing paper [2011]) might overlook an important and somewhat under-examined alternative (or middle path) which—I want to suggest—is opened by the synoptic philosophy of Wilfrid Sellars: such is the strategy that pursues the difficult but desirable goal of a fusion of the Kantian normative account of thought and scientific naturalism. This approach focuses attention on the ways in which scientific theorizing and practice are imbued with the conceptual, whilst offering an account of how such a normative sphere of rational agency is always to be ontologically located within a monistic and physicalist worldview, populated exclusively by the mind-independent entities (or processes) described by the natural sciences.

Ginev's stance aims to defend the rationality and the cognitive autonomy of science while rejecting any form of post-Cartesian epistemic foundationalism and representationalism hypostatizing the problematic notion of a subject (mind) – object (world) relation. The scientific everydayness of inter-related research practices and readable technologies, homologous to Kuhnian normal science, is a *distinguished* and *specified* epistemic enterprise yet not the *independent* activity of a mind oriented towards a world, inasmuch as it is a mode of cognitive engagement always already grounded upon a primordial interpretive fore-structure of Being by a situated subject.

Most notable, from the realist perspective through which I approach Ginev's work, is the rejection of the notion of "objective reality" over and above the world disclosed in the interpretative practices of scientific everydayness: the practical ways of being in the world of the scientific actor (as a mode of *Dasein*) precede and constitute the possibility for specific (meaningful) cognitive engagements with the world. This is in line with the post-Kantian, Heideggerian, reversal of epistemological priorities: only from within an always already present (ontologically prior) horizon of unreflective engagement with *zuhanden* tools encountered in their holistic "equipmental contexture" can we develop a theoretical, disengaged representation of *vorhanden* beings. Knowledge of the in-itself is constituted (made at all possible) by *Dasein*'s proper mode of primordial engagement with its surroundings, no matter how methodologically refined and technologically mediated such an engagement is. Ontic knowledge of entities is always dependent upon a prior ontological knowledge (even though the term "knowledge", as traditional goal of epistemic practices implying a subject-object relation between a knower and a known, should be traded for "understanding" [*Verstehen*]) of Being. A prior ontological knowledge which is not quite *a priori* in the strict Kantian sense: Heidegger's hermeneutic revision of Kantian transcendentalism discards the notion of necessary and universal transcendental conditions and recasts our inescapable finitude in terms of factical, contingent thrownness in a world—"Preliminary" [*Vorgängige*] would be a more authentically Heideggerian term. The propositional intentionality associated with the ontology of *vorhanden* objects is derivative from a more primordial immersion in a meaningful world of referential relations. Heidegger's well-known reinterpretation of Kant's project starts from the assumption that "[t]ranscendental philosophy denotes nothing but ontology": he insists that "ontology as science of being must be the science of the being-known of objects and of their possibility" and thus concludes that "[t]he interpretation of Kant's Critique of Pure Reason as epistemology completely misses the true meaning" (Heidegger, 1982, p. 128). Further, "Beings are in no way accessible without an antecedent understanding of being. This is the only possible meaning of Kant's thesis, which is frequently misunderstood and which is called his Copernican revolution" (Heidegger, 1997, p. 38).

My critique is relatively straightforward: Ginev's indirect endorsement of the Heideggerian hermeneutic *Wiederholung* of Kantian transcendentalism needs not be the *only* way to pursue the (almost) same anti-foundationalist agenda that guides him. The placement of minds squarely *in* and as *part of* the world, and hence the avoidance of the mystifying subject-world dualism, can take a naturalist rather than a hermeneutic slant without subscribing to empiricist dogmas. Ontological univocity can be naturalistically defended while preserving a reformed Kantian project distinct from the full-blown Heideggerian erasure of the interplay between the formally independent realms of rational thought and causally efficacious reality. The outcome is a "naturalism with a normative turn"—to use James O'Shea's (2007) incisive phrase—inspired by the work of Wilfrid Sellars, arguably capable of accounting for the continuity between normative-discursive practices and their natural grounds in a more successful way than Rouse's unpersuasive—in Ginev's account—naturalism (my interpretation of Sellars's work is here deeply influenced by O'Shea's compromise or "centrist" reading of Sellars's work, because it seems to me to be the most plausible and philosophically interesting blend between the "right-wing" or scientific reductionist interpretation and the "left-wing" or neo-pragmatist, inferentialist one).

Sellars's epistemological project offers the resources to see how there can be no foundational bedrock of independently acquired knowledge but, unlike Heidegger's equipmental holism, the dependence in question is from the rule-governed whole of our conceptual arsenal, not from the practical, concernful and careful engagement with our *Umwelt*: conceptualization is indeed what allows us to *interrupt and revise* our pragmatic routines, characterizing us as sapient reasoners rather than mere sentient organisms responding to the environment. The Sellarsian rejection of the "Myth of the Given"—the belief in any allegedly immediate, self-authorizing and epistemically foundational inner states—conjoined to his pragmatist insistence on the epistemic priority of linguistic performances within rational communities, allows us to reject Cartesian solipsism without opting for the phenomenological-hermeneutic stance of an inescapable ontological horizon comprising the totality of practices within which alone meaning can be discerned.

Sellars rejects the idea that natural kinds or particulars are neatly given to us in an experience-friendly taxonomical structure, that is to say, that our mind are innately (or divinely) furnished with a complete conceptual toolbox which happens to match experiential content (following in the footsteps of Kant's rejection of the pre-critical "theocentric paradigm" of cognition [See Allisson, 2004]). Indeed, Sellars agrees with Heidegger that experience-mediating concepts are, in a sense, *constituted* by a pre-theoretical purpose-driven comportment, a background of implicit familiarity, but for Sellars it is a functional, rule-governed constitution operated via the constant pragmatic and linguistic interactions with the world and with other sapient agents. The opposition between the two stances can be highlighted by considering how the later Heidegger would explicitly articulate his claim that beings are only accessible through a prior understanding of Being in terms of the power of language, and claim that

> [o]nly where the word for the thing has been found is the thing a thing. Only thus is it. Accordingly we must stress as follows: no thing is where the word, that is, the name is lacking. The word alone gives being to the thing. (Heidegger, 1971, p. 62)

Surely, we can avoid interpreting such claims as openly endorsing linguistic idealism: following this passage Heidegger considers and concedes the language-independent existence of a techno-scientific artefact like the (then contemporary) Sputnik satellite. Yet he also proceeds to insist on its dependence on the world of linguistically articulated social practices and their world-ordering power, as the background of linguistic disclosure of Being that alone makes technoscience possible.

For Sellars, on the other hand, socially regulated inter-subjective language is a pre-requisite for thought episodes, and thus logically prior in the order of knowing, i.e. it has a formal autonomy which plays an epistemologically constituting role (as Sellars claims: "[t]he primary mode of being a thinking as far as we understand it and seek to know what it is, *is* actually using language" [Sellars, 2009, p. 122]). It is not, however, ontologically prior in the order of being, since our perceptions depend upon sensible impressions received from a non-conceptual, mind-independent Real, and our linguistic thoughts upon physiologically specifiable events (neurophysiological processes). Here the distance with hermeneutic fore-structure is evident, since for the latter there is no temporal priority in its primordiality—our factical understanding of Being it is an "always already" for finite, thrown, beings like us—while the Sellarsian narrative is an evolutionary one (as dramatized in the "myth of Jones"), where the contours of our finitude are physiologically specifiable. Sellars's psychological nominalism commits him to the thesis that knowledge, and indeed "all awareness of sorts, resemblances, facts, etc., in short, all awareness of abstract entities—indeed, all awareness even of particulars—is a linguistic affair" (1997, p. 63), all perceptual takings are characterized by a conceptually (linguistically) articulated "thisness", not unlike Heidegger's basic "as" structuring of understanding. Sellars indeed highlights the primarily practical and engaged function of perception when observing that: "after all, the point of perception is not to illuminate the structure of the world, it is to get us around—like the point of pain is to get our hands off of stoves" (2009, p. 409). Yet our formally autonomous linguistic/conceptual performances causally depend on functionally isomorphic materially real physiological processes—an instantiation of what Sellars calls the 'Janus-faced character of languagings as belonging to both the causal order and the order of reasons" (1996, p. 110)—just as manifest, everyday appearances depend on real physical processes impinging on our senses. "Can we not interpret the receptivity involved [in the sensuous awareness of an object] in terms of 'purely physical' states, and attribute to these the role of guiding conceptualization?" asks Sellars (1968, p. 16) rhetorically.

The path to skepticism is thus foreclosed by a transcendental realist commitment to material processes. In Sellars's linguistic and naturalized Kantianism, the being of entities does not depend on our conceptual access to them, even while it is only through concepts qua rationally warranted rules of behavior and inference that we can gain any epistemic purchase on them. So there is indeed a constitutive fore-structuring of, or constitutive framework for, scientific research (or of any other form of experience), but it is a *conceptual-normative* rather than a *phenomenological-hermeneutic* one. The former, it seems to me, is a stance more authentically Kantian than the latter, where phenomenological anthropocentrism—the priority of *Dasein*'s existential analytic—trumps epistemic accountability. And Heidegger would happily concede this, considering that in his reckoning "[v]iewed as a *quaestio juris*, the transcendental deduction is the most disastrous segment of teaching in Kantian philosophy to which one can refer" (Heidegger, 1997, p. 209). Here the difference emerges clearly: for Heidegger the post-Critical enterprise is not a matter of epistemic rights, but of ontological affordances. As he continues later: "precisely not a *quaestio juris* but a *quaestio facti* lies at the center of the problem of the transcendental deduction. At stake here, of course, is not a factum in the sense of empirically determinable data and properties, but a fact in the sense of the ontological and essential structure [*Wesensbestand*] of *Dasein*, the transcendental constitution of the subject" (Heidegger, 1997, p. 224).

When I used in the foregoing considerations the term "phenomenological anthropocentrism", it was not my intention to involve in this qualification the sense of quasi-divine act of *Weltbildung* on *Dasein*'s part. "Phenomenological anthropocentrism" is to be applied to existential hermeneutics insofar as the always-already disclosedness of the world is only revealed through *Dasein*'s conceptually inexpressible *Befindlichkeit*. Even though the world is not a necessary production *of* the subject, it is primarily something that factically happens *to* the subject in his experiential domain and habitual dealings (Sellars makes the point that "[t]o say that man is a rational animal, is to say that man is a creature not of *habits*, but of *rules*" [1980, p. 298]). The anti-realist—strong correlationist, in Quentin Meillassoux's (2008) useful term—outcome of Heidegger's revised Kantianism is, somewhat ironically, a product of Heidegger's distancing from post-Kantian idealism. Heidegger's rejection of the B edition revision of the Metaphysical Deduction in the Critique of Pure reason is motivated by the excessive role, in his eyes, that the pure unity of apperception and the categories come to have *vis-à-vis* the three temporal syntheses, and the consequent demoting of imagination as a subordinate moment of the understanding. In Heidegger's view this shift in focus (roughly, that from a finite subject thrown in time to a formal transcendental subject) is the root, considering the problems of grounding that Kant's transcendentalism left unresolved, of the German idealist reaction. Heidegger indeed asks: "[w]hat is the significance of the struggle initiated in German idealism against the 'thing in itself' except a growing forgetfulness of what Kant had won, namely, the knowledge that the intrinsic possibility and necessity of metaphysics, i.e., its essence, are, at bottom, sustained and maintained by the original development and searching study of the problem of finitude?" (1962, p. 252–253). Heidegger's firmness on describing a pre-conceptual, receptive 'openness' rather than on the articulation of normative/structuring role of transcendental formal conditions (a receptivity which is the mark of finitude since "[f]initude consists in the reception of that which offers itself" [1962, p. 153]) is thus meant to save the independence of the object in itself, at the price, however, of making the event of its disclosure quite arbitrary and impervious to any attempt of discursive and inter-subjective epistemological justification. His insistence (especially in his later work) on *Dasein*'s passive receptivity to a meaning-bestowing event of world-disclosure, then, seems to take the form of a particular, historicized form of Sellars's myth of the categorical given, the belief in the givenness to the subject of an already-categorized world. This is the nostalgia of Being which Heidegger ultimately succumbs to, and the critique of which will turn out to be one of the founding gestures of Derrida's project, thus readable as itself a rejection of Heidegger's peculiar "myth" of the givenness of Being.

In rejecting Cartesian foundationalism and representationalism, then, we can part ways with the Heideggerian hermeneutic project and renovate a Kantian *quid juris* transcendental standpoint. Sellarsian Kantianism emphasizes the necessity of *intersubjective* epistemic accountability (one's belonging to the space of reasons) while Heidegger's Kantianism is more concerned with an anonymous and *trans-subjective* horizonal constitution, *Dasein*'s situational fore-structuring, which precedes (transcends) all context-specific practices of rational justification and is grounded in an unthematizable *Jemeinigkeit.* While Robert Brandom's socio-pragmatist reconstruction of Heidegger (2002, Ch. 10), paints a portrait of Heidegger with a Sellarsian brush which would seem to undermine my opposition between the two, this is done at the cost of presenting an ontologically anodyne Heidegger, profoundly at odds with Heidegger's later work. I thus agree with Ginev, when he claims that "misinterpreting the *Zuhandensein* as the realm of cultural values and significances, which has an autonomous being in itself, Brandom is blurring the essential differences between neo-pragmatism and hermeneutic phenomenology" (2011, p. 52–53). Ginev is right here, to the extent that, according to the hermeneutic stance, our social practices and interactions with tools are preceded by an indispensable pre-theoretical understanding of Being. However, the "genuine" (non Brandomian) Sellarsian position allows us to supplement social pragmatism with a robust form of scientific realism (the aspect of Sellars's thought Brandom is indeed less keen on), offering a mind- and society- independent scientific ontology as ultimate ground of possibility to which the manifest image of persons (qua members of a rational community sharing goals and values) and their intentional stances can be joined.

For Sellars the structure of intentionality can be naturalized by deriving it from concepts pertaining to overt linguistic behavior, not referring to any real property of thoughts as items in the real order (see Sellars, 1997, pp. 97–98). Deflating the notion of thought from an ontological abstract entity to a functional kind modelled on the overt use of language, Sellars avoids the notorious epistemological conundrums of a problematic semantic relationship between thought and world. To admit, with Sellars, the epistemological primacy (the always-alreadyness) of our immersion in linguistic and non-linguistic practices (responses, inferences, and actions) does not force upon us the ontological conclusion that our thoughts and the objects which they purport to describe do not enjoy a material (non-intentional) existence, and that a generalized notion of intentionality depends on a pre-conceptual understanding of Being. These are two distinct post-Kantian options, and more precisely two different interpretations of the Kantian theme of our discursive finitude. For Heidegger the (in-itself) Being of entities is inseparable from the conditions of their disclosure, as revealed through *Dasein*'s concernful dealings. For Sellars the mind is indeed active, insofar as the condition of possibility of experience of objects *as* objects is the possession of a holistic battery of socio-linguistically articulated concepts, so that "a diversified conceptual framework is a whole which, however sketchy, is prior to its parts, and cannot be construed as a coming together of parts which are already conceptual in character" (Sellars, 1963, p. 6). However, our conceptual access to objects does not coincide with, nor determine their real material being: Sellars observes that, in events of conceptual revision, "we do not, *eo ipso*, change that to which we are responding. [That] [t]here are items . . . to which we respond is a dimension of givenness (or takenness) which is not in dispute" (Sellars, 1981, p. 20). Sellars's naturalism aims to deliver an ontologically monist theory of representation (his theory of "picturing") wherein the relationship between normatively organized thoughts/languagings and the objects they stand in relation to is resolved in a structural isomorphism or correspondence between material items in the real order (such as neurophysiological states and external material phenomena): "[t]he fundamental job of singular first-level matter-of-factual statements is to picture, and hence the fundamental job of referring expressions is to be correlated *as simple linguistic objects* by matter-of-factual relations with single nonlinguistic objects" (Sellars, 1968, p. 124).

Rorty commented that "[p]icturing is for Sellars what disclosedness is for Heidegger. It is the extra dimension which relates social practises to something beyond themselves. . . . In Sellars's case this non-human something is 'the world'. In Heidegger's case it is 'Being'" (1991, p. 152). Sellars's goal is that of a thorough categorical *reconceptualization* of the manifest world, so that it would in principle be possible to perceive the world *directly* in scientific-theoretical terms (non-inferentially perceiving unobservables).[1] This clashes directly with Ginev's account of the visualizability ("first and foremost a phenomenological concept" [2011, p. 27]) of theoretical objects as contextual constitution of a meaning based upon the commerce with readable technologies, such objects acquiring existence only within the horizon of an interpretive scientific community: in the former case, the conceptual framework of the manifest image is abandoned or reshaped under the pressure of the constituents of the real world; in the latter, such unobservables are constituted by the community's textual-scientific practices, so that "[b]y appropriating horizons of possibilities on which theoretical objects are inscribed, one is transferring the latter to the realm of entities that are ready-to-hand" (Ginev, 2011, p. 29). Bluntly put, the Sellarsian realist bends the manifest image in order to accommodate (and to be able to "see") theoretical entities; the hermeneutic realist bends the theoretical entity in order to bring it down to the "seeing" or "appearing" range of the manifest everydayness of the *Lebenswelt*.

More importantly, to my reckoning, such a revised (naturalized) Kantian view of rule-governed conceptual systems, transcendentally determining the experience of objects, can be interpreted as allowing its own critical revision in the face of empirical experimentation, for the latter could eventually reveal (as it already has in the past) that the basic concepts which we apply to our familiar world fail to find any referent in the reality discovered through scientific inquiry. Furthermore, Sellars's local holism regarding concepts allows us to make sense of a plurality of conceptual frameworks synchronically possessed by a same subject,[2] and to account for the gradual process of the shift in meaning of our concepts, considering how such a meaning is not dependent on a primordial experiential correlation but on the pragmatics of inferential use (the latter being the core tenet of the philosophical tradition of inferentialism [see Brandom 2000]. For an excellent history and pre-history of inferentialism, see Peregrin [2013]). Sellars articulates this "pragmatic conception of the a priori" by explaining that:

> I see nothing horrendous in the notion that a language or conceptual framework brings with it a commitment to certain logically synthetic propositions, provided that it is recognized that there is more than one pebble on the beach, i.e. that there are many alternative frameworks, one of which the world persuades us to adopt (or, better, adumbrate), only to persuade us later to abandon it for another. (Sellars 1963, 293)

On the other hand, it is not clear how hermeneutically inflected transcendentalism can accommodate any plurality of conceptual frameworks and the existence of dynamical or relativized a priori principles (Friedman, 2001), rooted as it is on the notion of a condition of knowledge which is at the same time the condition for the disclosure of beings themselves. The hermeneutic paradigm, wherein a fore-structure of constitution is always-already conditioning the subject's epistemic access to the world, entails impersonal, anonymous conceptual schemes which are immune from revision and correction in the face of any evidence or "adumbration" from the empirical world (for the simple reason that no such evidence could be given outside of these very forms of givenness), and are merely to be passively accepted. New interpretive possibilities engendered by the interrelatedness of practices remain bounded by the general horizon created by a particular understanding of Being and, more generally, are a consequence of our existential finitude. No matter how "potentially infinite" the set of possible representings of theoretical objects is, the horizon of possibilities remains hermeneutically pre-constituted, and no amount of ontic interaction with the techno-scientific environment can lead us to radically revise such ontological pre-understanding.

If a bit of word-play can be forgiven (and whilst discussing Heidegger I would say that it can) the opposition between the two stances can be made evident by examining the role of the term "manifest". For Heidegger, ontico-scientific representation of an entity is possible only after it is made ontologically manifest as belonging to a meaningful network of intentional human concerns: "I can only objectify something if this something has beforehand presented itself to me as manifest [*Etwas vergegenständlichen kann ich nur, wenn dieses Etwas mir zuvor schon als Offenbares vorliegt*]" (Heidegger, 1996, p. 223; my Translation). The term "manifest" here is employed as translating the German *Offenbares*, or "revealed" i.e. that which is disclosed by a deep understanding of Being. For Sellars that which pertains to the "manifest image" similarly belongs to the conceptual framework which puts persons and their intentional stances on the foreground, the "framework in terms of which, to use an existentialist turn of phrase, man first encountered himself—which is, of course, when he came to be man" (Sellars 1963, p. 6), but the ontological priority is reversed: manifest entities and the normative conceptual space they belong to are both (logically) non-reducible and (ontologically) reducible to non-manifest, scientific ones. Most of that which is manifest, for Sellars, will be, in a Peircean regulative ideal of completed science, ultimately replaced according to the ontology of fundamental science, as opposed to the manifest entities which, for Heidegger, *cannot but be* the basic building blocks of our proximal environment and that only by a conscious effort of "de-worlding" can be seen as objects amenable to scientific description.

The hermeneutic realist notion of transcendence of a *world* "devoid of essence behind or beyond the interrelatedness of practices" (Ginev, 2011, p. xxiii), where objects of inquiry are meaningfully co-constituted within the world of scientific practices might be still robust enough, as Ginev intends, to reject social-constructivist relativism (and its reversed essentialism), but far too weak to offer an account of conceptual progress in terms of increased purchase upon *reality*. Sellars contends that

> [i]n ancient time, changes in the rules of language were very slow. Man was content to be baffled. Since the birth of modern science, man has constantly remodeled his language [i.e. his or hers conceptual framework]; indeed, from the standpoint of the anthropologist, science consists exactly in the attempt to develop a system of rule-governed behavior which will adjust the human organism to the environment. If there are regularities in the world, it is only by means of regularities in behavior that we can adjust to them. This process of adjustment can be speeded up by the deliberate exploration of alternative linguistic structures. (Sellars, 1980, p. 152).

It seems to me that the phenomenological-hermeneutic stance, presuming our relation with the environment to be one of ontological interpretive disclosure rather than rule-governed adjustment, one where our finitude entraps us in a mutual interdependence with a world-disclosing event of Being through which alone our coordinates of thought are bestowed, is intrinsically pervaded by a certain romantic taste for bafflement. The rationalist and "disenchanted" Sellarsian approach, then, turns out to be *more* open to the possibility of conceptual innovation through the collective employment of self-correcting scientific creativity than those philosophies of nature that, after Heidegger, vouch for a pre-conceptual exchange or "primordial dialogue" with (and in general the experiential immediacy of) a natural world always-already pregnant with meanings and purposes.

This comes to have bearing on Ginev's interest in presenting a socially progressive picture of science. His phenomenological-hermeneutic commitments make his notion of the progress of science, even after his attempts to defend its cognitive specificity, too anodyne and contextual for the realist's taste, lacking the explanatory power granted by reference to mind-independent entities. Ginev explains that: "[f]or hermeneutic realists, the aim of science is to reveal as much as possible the potentiality-for-being of domains of doing research (i.e., to reveal possibilities that are forgotten, hidden, or neglected in domains' normal science)" (Ginev, 2011, p. 165, n.11). But the everydayness of normal science, and its constitutive conceptual frameworks, needs to be continuously broken and reassembled through deliberate cognitive efforts which cannot be limited to moves within an unbreakable circle of readable technologies and representings. (I should note that Ginev's account—influenced by Hans-Jörg Rheinberger's work—relies on an interpretation of Derridean textual deconstruction according to which there is *literally* nothing outside the (scientific) text and thus reducing the scientific object to a bundle of technologically-produced inscriptions, which I consider an objectionable (yet all too common) conflation of Derrida's theses with linguistic idealism).

Even while rejecting social constructivist and postmodern sceptical dissolutions of "nature", Ginev, by endorsing the Heideggerian hermeneutic-ontological revision of the transcendental project, homogenises into an as omnipervasive as unthematizable "context of constitution" two heterogeneous tasks, the *scientific* and the *philosophical*: the former being a first-order experimental inquiry into the mind-independent natural world guided by own internal rules (theories, experimental methods, etc.), the latter the effort to evaluate, modify, and create conceptual frameworks (including the revision of our realist commitments to theoretical objects) which will function as historically mutable, yet *progressive*, transcendental coordinates for the practice of science itself. (It needs mentioning that the difference I am insisting upon here has direct consequences on the relevant concept of truth, scientific or not. While Sellars does consider truth as semantic assertability relative to conceptual systems, regulating practices in both the scientific and everyday settings, he considers some such systems as more adequate than others, where atomic matter-of-factual statements better *picture* states of affair in the world. Heideggerian truth completely forfeits adequation or correspondence in favor of a form of epochal disclosure which is as unrevisable as it is arbitrary). As Michael Friedman, who most forcefully articulated such a neo-Kantian approach to the philosophy of science, put it,

> [s]cience, if it is to continue to progress through revolutions, … needs a source of new ideas, alternative programs, and expanded possibilities that is not itself scientific in the same sense —that does not, as do the sciences themselves, operate within a generally agreed upon framework of taken for granted rules. For what is needed here is precisely the creation and stimulation of new frameworks or paradigms, together with what we might call meta-frameworks or meta-paradigms – new conceptions of what a coherent rational understanding of nature might amount to – capable of motivating and sustaining the revolutionary transition to a new first-level or scientific paradigm. (Friedman, 2001, p. 23)

I want to touch a final metaphilosophical point, which I consider as the most vital and radical node of our my disagreement with Ginev: Sellars famously described his project as an attempt to move (analytic) philosophy from its Humean (empiricist) phase towards a belated Kantian (normative-inferentialist) one.[3] Ginev's project amounts to a similar reaction to the foundationalist-empiricist (mostly Cartesian) doxa he sees as still rampant in the philosophy of science, and a move towards a Heideggerian (hermeneutic-phenomenological) phase. The former path eschews epistemic fundamentalism but intends to preserve a naturalism-friendly transcendental project where norm-bound agents discover their ontological ground in terms of a purely causal universe; the latter path tends to reject, with fundamentalism, the very necessity of epistemology, and favors (or in fact it *asserts*) the primacy of a hermeneutic existential co-implication of interpreter and interpreted (c.f. Rorty's jump from his "left-wing" reading of Sellarsian anti-foundationalism to his idiosyncratic Heideggerian/Derridean sceptical linguistic idealism). So Ginev's contention is that the phenomenological-hermeneutic methods of his cognitive existentialist interpretation of scientific practice offer the "*only* possible access to the whole of nature" (Ginev, 2011, p. xxxiv. Emphasis added).

Ginev, to his signal merit, strongly insists on the *political* necessity of the cognitive specificity of the scientific enterprise (and is well aware of the ironist lures of postmodern sirens) as he stresses that it is precisely the study of the hermeneutic fore-structure of scientific research which helps justifying it. However, I remain unconvinced about the social strength of the cognitive autonomy of a discipline not grounded upon the ontological autonomy of its object matter, leaving its metaphysical commitments *and* ethical compass all too malleable. Similarly, I am unpersuaded by Ginev's warnings against the perils of "the self-imposed ideology of scientism" (Ginev, 2011, p. 145) for a democratic society. As is often the case, I am not sure what 'scientism' really amounts to: if it is the claim that science is the *only* and *incorrigible* epistemic authority, I reject it as untenable. If, on the other hand, it is the claim that science, in virtue of its metaphysical realist credentials offering rationally justified evidential grounds, should serve as ultimate judiciary court of appeal (but note, not directly holding *legislative* powers) used to resolve those issues that our society is unable to settle through a Marcusian "dialogical" approach with nature vouched for by Ginev, I am happy to endorse scientism. Ginev's characterization of scientism as a project of "complete technological manipulation and instrumental transformation of natural resources" (Ginev, 2011, p. xxxiv) seems to me to be a quaint—and all too Heideggerian (and Marcusian)—caricature which can far too easily be strategically employed as a bogeyman by technophobe reactionary forces.

Am I delineating an unbridgeable gap between me and Ginev, a fundamental irreconcilability in philosophical temperament? This seems to me a pressing question, and my entire interpretation of Ginev's enterprise is orbiting it. I cannot do full justice to it in what remains of this paper, but it seems to me that this contrast goes deeper than the worn-out opposition between continental and analytic approaches, an opposition that (unlike Ginev who, even while confidently analyzing the work of several analytic philosophers, seems happy to identify his work as "continental") I would like to see disappear into a near-future synoptic fusion of the conceptual resources of the two traditions. It should be added that, adopting this familiar taxonomy, Ginev's "continental" affiliation could be seen as going somewhat counter-current, considering that, as I explained above, he fully endorses what has been recently become an object of vehement critique in those increasingly prominent quarters of continental philosophy currently endorsing a realist renaissance: a *correlationist* stance (following Quentin Meillassoux's coinage) which presumes an unavoidable correlation between subject (thought) and object (being). The strong variety of correlationism, typified by Heidegger's phenomenological ontology (and by extension, Ginev's cognitive existentialism) assumes, in opposition to both Kantian criticism and Hegelian idealism, the impossibility of knowing and of meaningfully conceiving of a mind-independent object, and stresses the facticity of the correlation between knower and known, a status of throwness predicated upon the existential finitude of a subject-within-a-world not of her making (for an overview of these realist positions, often associated with the term "speculative realism" see Gironi [2012b] and the papers in Volume IV of the journal *Speculations,* in particular Daniel Sacilotto's contribution, taking, like me here, a Sellarsian stance to the problem of representation).

The split I have in mind as underlying the disagreement I have with Ginev is that between, on the one hand, a philosophy giving methodological priority to the givenness of lived experience and the immediacy of phenomenological intuition and, on the other, a philosophical orientation insisting on the rationalist priority of the concept (and two different conceptions of subjectivity are crucially at stake in this division). Knox Peden, in an outstanding report on the under-explored intellectual landscape of mid-twentieth century French philosophy, eloquently articulates this conflict by describing it as a clash

> between a philosophy which emphasizes the limits of rational thought to the profit of a more primordial, ineffable experience or intuition, and a philosophy which insists upon the capacity of rationalism to transgress the limits of lived experience to articulate conceptual insights of a universal or indeed absolute variety. (2011, p. 365)

Peden follows lines famously drawn by Michel Foucault (in Canguilhem 1978) and rehearsed, more recently, by Alain Badiou (2005). However the point stands even when these national boundaries are left behind, and indeed the transversal nature of this scission seems to me to clearly indicate the artificial nature and limited usefulness of the "continental" and "analytic" labels (I therefore wish to avoid my critique to be trivialized as an "analytic" rebuttal to Ginev's "continental" philosophy of science).

This divergence of attitudes forces the adoption of a stance privileging either "a respect for the limits of reason in light of the constraints of finitude, or reason's capacity for transgressing these constraints in the production of 'true ideas'" (Peden, 2011, p. 388). This tension recapitulates a profound gap running throughout the entire history of philosophy, one across which, one foot precariously on each edge, towers the monumental figure of Kant. Hence explained my approach to Ginev's work in the key of two different reactions (Heidegger's and Sellars's) to the Kantian enterprise. My preference, it should be obvious by now, goes to the rationalist and conceptualist side, and yet I take the Sellarsian revision of Kant to be uniquely equipped to offer a naturalist worldview that, whilst firmly committed to the priority of rational reflection upon pre-conceptual lived experience and to the Peircean ideal of an as yet unfinished scientific image where persons as intentional wholes disappear from view, gives its due to the non-eliminable role of practical reason in regulating our embeddedness in a human community fraught with meanings and intentions (thus evading what Ginev calls the "depressing dilemma" [2011, p. xxxiv] between nature and culture). In phenomenological-hermeneutic terms, we can aim at "objectivized" and yet never fully "demundanized" subjects: that is the pay-out of Sellars's *synoptic vision*. As Willelm DeVries puts it: "Sellars essentially accepts the ontology of the reductivist and the ideology of the non-naturalist. We can recognize truths that cannot be expressed in the vocabulary of physicalistic science without accepting the existence of supernatural objects" (in DeVries, 2009, pp. 225–226). Unlike some hard-core naturalists, I do not covet the destruction of phenomenology, which I take to be an essential part of the philosophical enterprise (as did Sellars, envisioning it as part of the *philosophia perennis* reflecting upon that sophisticated common-sense framework which is the manifest image) as long as its focus on the intentional sphere of subjective experience is prevented from either replacing a conception of the subject as norm-bound rational agent (as in the case of Husserlian transcendentalism) or shifting from a methodological guideline to an ontological thesis (as in Heidegger's project of phenomenological ontology).

As much as I have admired Ginev's meticulous account of science's practices, and his careful and thorough engagement with the most prominent figures of contemporary philosophy of science, my suspension of disbelief in the phenomenological-hermeneutic axioms that his cognitive existentialism carries can only be pushed as far as an appreciation of its refined conceptual architectonics and its insistence on the practical dimensions of science (experimentation, calibration, measurement, inscription, modelling, etc.). It is an appreciation, however, that is not quite enough to persuade me to abandon my rationalist and realist commitments. I share, however, Ginev's broader goal, that of constructing a philosophical structure capable of assigning to the scientific enterprise a justified central and autonomous role in democratic society. Hopefully these disagreements will help shine some light on the way towards it.[4]

NOTES

1. See Brandom in Smith (2002) for a Sellarsian account of the observability of subatomic particles.
2. Brown (2007) significantly expands this Sellarsian insight, arguing for a plurality of "guiding assumptions": propositions treated as necessary truths *in a given time and domain*, expressing our present best view of the laws of nature, and requiting the acceptance of material rules of inference allowing moves between concepts.
3. This motto is reported by Rorty (in his "Introduction" to Sellars 1997) and Brandom (2000, 32).
4. I would like to thank Daniel Sacilotto for his helpful comments on a draft of this paper.

WORKS CITED

Allison, Henry. (2004) *Kant's Transcendental Idealism: An Interpretation and Defense*. New Haven and London: Yale University Press.

Badiou, Alain. (2005) "The Adventure of French Philosophy," *New Left Review*, 35, pp. 67–77.

Brandom, Robert B. (2000) *Articulating Reasons: An Introduction to Inferentialism*. Cambridge, Massachusetts and London: Harvard University Press.

————. (2002) *Tales of the Mighty Dead: Historical Essays in the Metaphysics of Intentionality*. Cambridge, Massachusetts and London: Harvard University Press.

Brown, Harold (2007) *Conceptual Systems*. Oxford and New York: Routledge.

Canguilhem, Georges. (1987) *The Normal and the Pathological*. Dordrecht: Reidel.

DeVries, Willem, ed. (2009) *Empiricism, Perceptual Knowledge, Normativity, and Realism. Essays on Wilfrid Sellars*. Oxford. Oxford University Press.

Friedman, Michael. (2001) *Dynamics of Reason*. Stanford: CSLI Publications.

Ginev, Dimitri. (2011) *The Tenets of Cognitive Existentialism*. Athens, Ohio: Ohio University Press.

Gironi, Fabio. (2012a) "A Hermeneutic Defence of the Cognitive Autonomy of Science," *Metascience*, 22:2, pp. 461–464

————. (2012b) "Between Naturalism and Rationalism: A New Realist Landscape," *Journal of Critical Realism*, 11:3 pp. 361–387.

Heidegger, Martin. (1962) *Kant and the Problem of Metaphysics*. Bloomington: Indiana University Press.

————. (1982) *The Basic Problems of Phenomenology*. Bloomington: Indiana University Press.

————. (1971) *On The Way to Language*. New York: Harper and Row.

————. (1996). *Einleitung in die Philosophie. Freiburger Vorlesung Wintersemester 1928/29.Gesamtausgabe, vol. 27*. Frankfurt: Vittorio Klostermann, 1996.

————. (1997). *Phenomenological Interpretation of Kant's Critique of Pure Reason*. Bloomington: Indiana University Press, 1997.

Meillassoux, Quentin. (2008) *After Finitude: An Essay on the Necessity of Contingency*. London: Continuum.

O'Shea, James. (2007) *Wilfrid Sellars: Naturalism with a Normative Turn*. Malden: Polity.

Peden, Knox. (2011) "Descartes, Spinoza, and the Impasse of French Philosophy: Ferdinand Alquié Versus Martial Gueroult," *Modern Intellectual History*, 8:2 pp. 361–390.

Peregrin, Jaroslav (2014) *Inferentialism (Why Rules Matter)*. London: Palgrave Macmillan.

Rorty, Richard. (1991) *Objectivity, Relativism, and Truth. Philosophical Papers*. Cambridge: Cambridge University Press.

Sankey, Howard, & Ginev, Dimitri. (2011) "The Scope and Multidimensionality of the Scientific Realism Debate," *Journal for General Philosophy of Science*, 42:2, pp. 263–283.
Sellars, Wilfrid. (1963) *Science, Perception, and Reality*. Atascadero: Ridgeview.
———. (1968) *Science and Metaphysics. Variations on Kantian Themes*. London and New York: Routledge and Kegan Paul.
———. (1980) *Pure Pragmatics and Possible Worlds*: The Early Essays of Wilfrid Sellars. Ed. By Jeffrey Sicha. Atascadero: Ridgeview.
———. (1980) *Naturalism and Ontology*. Atascadero: Ridgeview.
———. (1981) "The Lever of Archimedes," *The Monist*, 64:1, pp. 3–36.
———. (1997) *Empircism and the Philosophy of Mind.* With an Introduction by Richard Rorty and a study guide by Robert Brandom. Cambridge, Massachusetts: Harvard University Press.
———. (2009) *Notre Dame Lectures 1969–1986—The Bootleg Version*. Ed. and intro. by Pedro Amaral, 2009. Available on the Sellars website maintained by Andrew Chrucky at http://www.ditext.com/amaral/wsndl.pdf
Smith, Nicholas H. ed. (2002) *Reading McDowell: on Mind and World*. London and New York. Routledge.

Five

SCIENTIFIC PRACTICE AND MODES OF EPISTEMIC EXISTENCE

Jeff Kochan

1 Introduction

The methodological doctrine that science is best studied in terms of epistemic practices, rather than in terms of conceptual schemes, is now well-established and widely followed in the field of science studies. Indeed, over the past thirty years, the very idea of an epistemic practice has achieved the status of a basic concept within that now maturing field. What prompted this shift in perspective, and the enthusiasm with which the idea of practice was taken up by philosophers, sociologists and historians alike, is an as-yet unwritten chapter in the history of science studies as a discipline. Needless to say, we now live in an intellectual climate where a wide range of researchers employ the term 'practice' with the same unselfconscious ease which had once accompanied the terms 'conceptual scheme', 'mentality' and 'worldview' in studies of the scientific enterprise.

I specify scientific practices as *epistemic* practices in order to highlight the fact that practice-based explanations of science are still largely concerned with scientific *knowledge*. Epistemic practices are specifically knowledge-making practices, and they now serve as a basic analytical resource in the field of science studies. Where researchers had once focussed on systems of scientific concepts, and their relation to independently existing matters of natural fact, they are now increasingly focussed on the techniques by which concepts, and even matters of natural fact, are constructed through disciplined scientific activity. As Ian Hacking (1992, 1), a key early contributor to this methodological shift, aptly expressed the difference: "The focus is on how we find out, not on what we find out." The study of scientific knowledge is not exhausted by studies of what scientists know. It is now recognised that much importance also lies in studying how they know.

This distinction between the what and the how of scientific knowledge is closely related to a more well-known distinction between knowledge-that and knowledge-how, or what might also be termed a distinction between conceptual, or theoretical, knowledge, on the one hand, and practical knowledge, or know-how, on the other. This can be rendered more succinctly as the distinction between theory and practice. The point to remember, however, is that this is a distinction *within* the semantic space occupied by the term 'knowledge.' The study of scientific practice is part of the study of knowledge, but by stretching the semantics of 'knowledge' to also include epistemic practices, the meaning of 'knowledge' itself undergoes a significant change. Knowledge can no longer be viewed as something exhaustively expressible in a fact-stating language. The new idea is that scientific knowledge cannot be captured solely in terms of propositional structures. How we find out cannot be reduced to what we find out. Know-how, epistemic practice, has a life of its own, independent of propositionally structured theoretical knowledge. This insight has proven enormously fruitful in the study of scientific knowledge, fuelling a "practice turn" which has generated an impressive and influential body of work. With only slight exaggeration, one may say that the recognition of non-propositional, practical knowledge as a distinct and legitimate topic for research marked a key foundational moment in the now maturing academic discipline of science studies.

Yet, as might be predicted whenever exciting intellectual innovation combines with enthusiastic discipline building, practice-based theorists of science have sometimes been too eager to assert the novelty of their position at the expense of nuance in their analyses. I wish to focus here on one particular, consequential instance where intellectual nuance has been compromised by enthusiastic discipline-building in practice-based studies of science. Two champions of practice, Dimitri Ginev and Joseph Rouse, base their accounts on a critical appropriation of the earlier work of Martin Heidegger. This appropriation was a "critical" one because they both argue that Heidegger, whilst identifying and emphasising the distinction between theory and practice in studies of science, lacked the nerve to follow through with this insight by arguing that scientific theory must be explained reductively in terms of scientific practice. In their view, Heidegger initiated the practice turn in studies of science, but he did not himself turn far enough. To turn "far enough" would mean to have turned entirely away from a conception of scientific knowledge in terms of theory, and so entirely and exclusively towards a conception of scientific knowledge in terms of practice. In the disciplinary history of science studies, Heidegger can thus only be considered a sub-hero, because he did not pledge his unmitigated and unequivocal allegiance to the flag of epistemic practice.

My goal here is neither to judge degrees of heroism nor to patrol disciplinary boundaries. I would like, instead, to introduce some nuance into the now-fading novelty of practice-based studies of science. I will do this by offering a different interpretation of Heidegger's account of scientific practice and its relation to scientific theory. In my view, the reason Heidegger did not collapse theory into practice was not because he lacked intellectual nerve, but because he rejected the idea that either theory or practice could be entirely collapsed one into the other. For Heidegger, the choice between theory and practice was not forced by a distinction between two equally fundamental and mutually exclusive phenomena. It was rather a choice between two distinct epistemic modes of the same underlying phenomenon. That underlying phenomenon was, for Heidegger, *existence*. As I will present him, Heidegger took theoretical and practical knowledge to be different determinations within the same experiential field of scientific existence.

I will first outline my own interpretation of Heidegger's existential account of science. In my view, the basic concept of Heidegger's account was not practice, but existence. Heidegger describes the structure of modern scientific existence in terms of the "mathematical projection of nature." I then turn to Rouse's criticism of Heidegger's concept of mathematical projection. According to Rouse, the centrality of this concept in Heidegger's work demonstrates Heidegger's continuing reliance on a theory-based account of science. Ginev largely accepts Rouse's interpretation, and furthermore charges Heidegger with the intellectual sin of "mathematical essentialism." I finally move to deflect Rouse's and Ginev's criticisms of Heidegger on the basis of my own interpretation. I argue that Heidegger regarded the mathematical project as the existential ground state for modern scientific experience, a ground state which enables scientific experience in terms of both theory and practice. In opposition to Ginev, I argue that Heidegger's concept of the mathematical projection of nature should more properly called a doctrine of "mathematical existentialism."

2 Heidegger's Existential Conception of Science

In *Being and Time*, Heidegger (1962, 408) distinguishes between a "logical" and an "existential" account of science. The logical account understands science in terms of the representations of nature it produces, representations which have the structure of a coherent body of interconnected true propositions. On this account, in other words, science is viewed as a conceptual scheme. On the existential account, by contrast, science is viewed in terms of existence. It is a way of being-in-the-world, which brings to light natural entities as new objects for theoretical understanding. The existential account is meant to explain the emergence in experience of the logical account: it attempts to describe the existential conditions for the kind of epistemic experience presupposed by the logical account. The analysis is thus phenomenological in its method. The logical conception of science relies on a model of epistemic experience, of "mind" or "reason," as possessing a propositional structure. Heidegger argues that a phenomenological investigation of the "logical mind," as the cognitive vehicle for science as conceptual scheme, leads back to the existential conditions which ultimately allow for, and enable, this specific kind of epistemic experience.

It is important to emphasise that Heidegger does not challenge the legitimacy of the logical account as a tool for the analysis of science. He only challenges the legitimacy of claims regarding the fundamentality of that analysis. Treating reason as propositionally structured, and science as a conceptual scheme, may prove useful for some types of analysis, but this utility should not be misunderstood as a sign of fundamentality. A phenomenological investigation of first-person theoretical experience reveals it to be an only derivative form of epistemic experience. Heidegger is here grappling with the problem of grounding abstract, theoretical reasoning in the concrete materiality of human existence. His strategy for connecting the two is to shift the focus of analysis from theory to practice. He focuses on the concrete activity of scientific work as being-in-the-world, and attempts to show such concrete activities come to constitute, and to maintain, theoretical knowledge of nature.

Determining the conditions of possibility for the theoretical attitude involves an analysis of the existential conditions under which propositionally structured reason emerges from out of everyday, engaged activity in an environment. Heidegger (1962, 88-89) analyses this emergence in terms of four experiential steps. This four-step process may result from a sudden disturbance in the flow of everyday, engaged activity, as, for example, when the tip breaks off a pencil in the midst of writing, or a tire abruptly goes flat in the midst of cycling. The process may also spring from a more deliberate disengagement from an activity, as, for example, when one is "taking a rest" (Heidegger ,1962: 83). In the first step of the process, one holds back from immersed involvement so as to merely look, and no more, at the thing with which one had been involved. In the second step, pure looking becomes a thematising in which things are encountered as objects of perception. In the third step, perception interprets things so as to determine their properties. In the fourth step, determinate perceptual objects become the subject matter for propositional knowledge claims. Heidegger calls this four-step transformation in the way things are experienced a "change-over" in epistemic experience. It is the movement from non-deliberative, non-propositional know-how to deliberative, propositionally structured knowledge-that, from practice to theory as two distinct experiential modes of epistemic existence.

The change-over thus marks a shift in the existential structure of epistemic experience, a structure which Heidegger (1962, pp. 184–185) calls "projection." As the basic structure of epistemic experience, the projection is present even when no change-over has occurred. It resides just as much in the undisturbed practical activities characteristic of immersed involvement in a work-world, as in the deliberative and disengaged thematisation of a particular thing as the subject matter for a propositionally structured analysis.

Heidegger argues that scientific understanding is structured on the existential level by a particular kind of projection. The existentially decisive feature of science is, in his view, neither pure, unstructured observation, nor formal mathematical modelling, but "*the way in which Nature herself is mathematically projected*" (Heidegger, 1962, pp. 413–414). This mathematical projection determines the range of possible ways by which the modern scientist comes to make sense of nature, both at practical and at theoretical levels of experience. It furthermore serves as the existential source from which the logical conception of science draws its own currency. The change-over from immersed involvement to a theoretical conception of nature is a shift in the existential structure of the mathematical projection as the fundamental ground of scientific experience. According to Heidegger, within the range of epistemic experience enabled by this projection, pure looking shifts specifically toward a perception of things as objects quantitatively determinable in terms of such general categories as motion, force, location, and time. In other words, the projection enables an experience of things as objects with properties amenable to standardised forms of numerical measurement (Heidegger 1962, p. 414). Hence, Heidegger argues that the existential possibility of empirico-mathematical science is conditioned by the projection of nature as *a priori* measurable. Under such conditions, the existential horizon within which things can be experienced as scientifically intelligible simultaneously restricts the possible ways in which an entity may be thematically discovered as an objective feature or constituent of nature. Heidegger (1962, p. 414) writes that the aim of thematising is "to free the entities we encounter within-the-world, and to free them in such a way that they can 'throw themselves against' a pure discovering – that is, that they can become 'Objects.'" The mathematical projection thus provides the basic existential structure for the process of scientific discovery. It disposes scientists to develop and deploy specific investigative techniques and tools, in the very design of which the generalised, uniform measurability of natural phenomena has been taken for granted. Within the existential space opened up by the mathematical projection, these measuring techniques and tools eventually become normalised into something like what Hacking (1982, p. 73) has called a "regular technology."

Heidegger's existential account of science can thus be separated into two distinct components. The first component is his shift of focus from science as a conceptual scheme to science as a form of existence. The second component is his phenomenological description of the change-over in scientific experience and its relation to the mathematical projection. With this distinction in mind, let us now turn to the critical interpretation of Heidegger's work presented by two champions of practice-based science studies, Joseph Rouse and Dimitri Ginev.

3 Mathematical Essentialism: The Unnuanced Reading of Heidegger

In his influential 1987 book, *Knowledge and Power*, Rouse laid out the basic foundations for a "practical hermeneutics of science" based, in significant part, on a critical interpretation of Heidegger. The main shortcoming of Rouse's critical interpretation is its failure to recognise the two distinct components of Heidegger's conception of science, as outlined above at the end of the last section. Instead, Rouse mixes them together: on the one hand, embracing Heidegger's move away from the logical conception of science as a conceptual scheme, and, on the other hand, reading Heidegger as attempting to replace the logical conception with a conception of science as a material configuration of practice. The key interpretive infelicity in Rouse's reading of Heidegger is his identification of existence with practice, an identification which is absent in Heidegger's own work.[1] As a consequence, he cannot help but interpret Heidegger's discussion of the change-over as a regressive shift back from praxical existence to science construed as conceptual scheme (Rouse 1987, p. 103). Because Heidegger does not surrender everything to practice, his heart must still lie with theory, or so Rouse thinks. Ginev (2005, p. 199; 2011, p. 5, 103) fully endorses Rouse's reading, and follows him down this errant interpretive path. (Apart from this crucial point of agreement, Ginev is far from being an uncritical disciple of Rouse: he has vigorously criticised Rouse's failure to acknowledge the cognitive specificity of science (Ginev, 2005). As will become clear in the conclusion of this paper, my focus on modes of epistemic existence can accommodate the idea of cognitive specificity and so is immune to this criticism.)

Rouse's attempt to identify existence with practice apparently follows from his implicit reduction of the field of enquiry to only two fundamental and mutually exclusive analytical categories: theory or practice. As a consequence, he deprives himself of the possibility to understand Heidegger's concept of existence as phenomenologically prior to both. This then leads him to misinterpret Heidegger's existential conception of science as equivocating between competing theory-based and practice-based accounts of science. Rouse traces the source of this alleged equivocation back to Heidegger's concept of the change-over, arguing that the change-over marks Heidegger's residual attachment to a theory-dominated view of science. He thus rejects the mathematical projection as a retrograde move which disastrously betrays Heidegger's original insight that scientific knowledge is constituted in configurations of material practice (Rouse, 1987, p, 74). Indeed, according to Rouse (1987, p. 79), "the theory-dominant perspective that Heidegger still retains [...] reduces experiment to a merely incidental practice in science." He thus concludes that, in Heidegger's view, material practice "neither has 'a life of its own' apart from theory nor makes a distinctive cognitive contribution" (Rouse, 1987, p. 98).

According to Rouse (1987, p. 75), "it is not at all clear in [Heidegger's] account *how* one can get from a breakdown of practical involvement to the theoretical attitude. But once this happens, the ordinary functional contextuality of things gets *replaced* by the 'mathematical projection of Nature.'" Yet the lack of clarity bothering Rouse is not Heidegger's own, but rather belongs to Rouse's unnuanced interpretation of Heidegger. Because he does not recognise that the mathematical projection provides the existential ground for the change-over from practice to theory, Rouse ends up erroneously equating the mathematical projection with the theoretical attitude. Accordingly, he argues that Heidegger identifies the mathematical projection with decontextualising theorising and also attributes to Heidegger the view that "[w]hen we understand *theorizing*, we have understood what is essential about science" (Rouse, 1987, p. 96).

This unnuanced and errant reading of Heidegger has been picked up, and expanded upon, by Dimitri Ginev in his development of a cognitive existentialist account of scientific practice. Ginev (2011, p. 11) argues that a revision of Heidegger's allegedly problematic concept of the mathematical projection "leads directly to the doctrine of cognitive existentialism" which Ginev promotes as a superior alternative. Like Rouse, Ginev (2011, p. 3) misinterprets the mathematical projection as a theoretical phenomenon, reducing it to the "projection of a mathematical structure (e.g., a system of differential equations)." According to Ginev (2011, p. 3), the mathematical projection "discloses a domain of empirical objects of inquiry through the theoretical models it imposes." It is, he argues, the vehicle for a "ready-made formalism" (Ginev 2001, p. 13). On this reading, Heidegger's mathematical projection is not a general form of existence on the basis of which scientists experience things as measurable. It is instead attributed with the much more determinate structure of, to use Ginev's example, a system of differential equations. It narrowly and determinately shapes scientific experience by imposing "ready-made" theoretical models on objects of inquiry, rather than by more generally disposing scientists to experience things as amenable to procedures of measurement. On the basis of this reading, Ginev attributes to Heidegger a doctrine of "mathematical essentialism."

This violent reading of Heidegger has the effect of pushing his concept of the mathematical projection into the pigeon hole of a theory-dominant account of science. For the champions of practice – who view the field of inquiry in black and white, "either/or" terms – this flawed reading then presents them with an immediate, if rather artificial, threat to their own methodological commitments. Ginev's response to this invented threat is to "revise" Heidegger's allegedly theoretical concept of mathematical projection, arguing instead that the "mathematical projection is no more than a particular practice (or probably, a cluster of practices)" (Ginev 2011, p. 11). Having first forced Heidegger's conception into the pigeon hole of theory, Ginev then claims to save the day by reclaiming it for the pigeon hole of practice. Meanwhile, Heidegger's actual account, which presents the mathematical projection as a form of existence prior to both scientific practice and theory, remains untouched, standing neglected on the side lines whilst the champions of practice struggle heroically to clear the field of the last phantasms of theory. It is a self-produced philosophical theatre in which the protagonists enthusiastically promote their own novelty at the cost of interpretive nuance.

4 Mathematical Existentialism: The Nuanced Reading of Heidegger

In his purported "revision" of Heidegger's concept of the mathematical projection of nature, Ginev (2011, p. 13) is especially concerned to demonstrate not only that projection is practice, but also that this projective practice "cannot be isolated from the everydayness of doing research." This latter claim is meant to block a construal of the projection as a practice of "deworlding" which facilitates the "disappearance of any situatedness (or place) in what becomes mathematically projected and disclosed for an empirical investigation" (Ginev, 2011, p. 4). This criticism recalls Rouse's allegation, noted above, that Heidegger meant the mathematical projection to replace the "ordinary functionality" of things. The general idea is that the projection serves to release scientific experience from local contexts of situated research activity. Hence, both Ginev and Rouse equate the mathematical projection with a form of decontextualised theorising. Here is the key passage from Heidegger (1962, p. 413) on which this criticism is based:

In our first description of the genesis of the theoretical attitude [...] we have made basic a way of theoretically grasping entities within-the-world – physical Nature – in which the modification of our understanding of Being is tantamount to a change-over. In the 'physical' assertion that 'the hammer is heavy' we *overlook* not only the tool-character of the entity we encounter, but also something that belongs to any ready-to-hand equipment: its place. Its place becomes a matter of indifference. This does not mean that [the thing] loses its 'location' altogether. But its place becomes a spatio-temporal position, a 'world-point', which is in no way distinguished from any other. [...] [T]he releasing from such environmental confinement belongs to the way one's understanding of Being has been modified.

According to Ginev, this is the *locus classicus* for Heidegger's belief that the mathematical projection involves a decontextualisation which tears things from their place through an act of theory. This is what Ginev means by the "deworlding" of a thing, the disappearance of its "[s]ituatedness [...] within the interrelatedness of contexts of equipment" (Ginev, 2011, p. 4). But Heidegger does not say, in this passage, that the situatedness, or place, of the thing disappears. He says that it is *overlooked*: "Its place becomes a matter of *indifference*." The thing continues to be situated, to inhabit a place, but the agent now views this as irrelevant for her understanding of the thing. What has changed is not the thing's situatedness in an environment, but the agent's epistemic experience of the thing in relation to its environment. Ginev misreads Heidegger as attributing "releasement from an environment" to the thing, when, in fact, Heidegger attributes it to the way the thing is understood, or experienced, by the agent. Clearly Heidegger cannot view the thing as having been "deworlded," because he still attributes to it a "world-point." His argument is that the epistemic agent has come to view the world of the thing as a uniform nexus of coordinated spatio-temporal points, rather than as a qualitatively differentiated environment of unique places. The thing's spatiality, in other words, is now experienced as amenable to quantitative acts of uniform measurement.

Heidegger explicitly links these considerations to the history of science, citing the rise of mathematical physics as a "classical example" of the process he is describing. What he has in mind is the radical transformation from the medieval Aristotelian understanding of nature to the early-modern Newtonian understanding of nature. In a 1934 lecture, Heidegger spells this difference out in substantial detail, focusing especially on the nature of motion. In the Aristotelian tradition, a body moves according to its nature (Heidegger, 1993, p. 284). An earthly body will move downward, towards the earth. A fiery body will move upward, towards the heavens. This is because the earthly has its place below, and the fiery has its place above. Each thing has its place in the cosmos according to its kind, and it naturally moves toward that place when separated from it. On this view, the cosmos is divided up into qualitatively distinct regions, and each thing, according to its nature, finds its natural place within one of those regions.

The Newtonian account of motion is profoundly different. Newton's first law was meant to apply to all things, everywhere. Where each body was once viewed as moving in accordance with its own unique and intrinsic nature, the movement of all bodies is now viewed as governed by a single, universal law: "Every body left to itself moves uniformly in a straight line" (Heidegger, 1993, p. 286). In order for this law to gain credibility as a fundamental description of experience, the qualitative difference between places must become a matter of indifference: such differences are no longer scientifically relevant. There is no specific place in the cosmos where a thing is meant to belong. As a consequence, the concept of "place," as such, also changes. A place no longer has a unique and intrinsic nature; it is now just one generic location in relation to other such locations, one world-point amongst other world-points. On this basis, motion comes to be understood in terms of measurable distances between discrete world-points. Motion is now *amount* of motion (Heidegger, 1993, pp. 287–288). The underlying experience of space has changed over from an Aristotelian one of differentiation and hierarchy within a cosmos to an early-modern Newtonian one of uniformity and equivalence within a universe. Where space was once understood in qualitative and differential terms, it is now understood quantitatively, as a uniform medium fundamentally amenable to standardised forms of measurement.

Ginev (2011, p. 7) argues that standardised practices of measurement give rise to formalised, geometrical theories of space. He thinks that this as an argument against Heidegger's conception of the mathematical projection of nature, because he misidentifies the projection with formal, mathematical theory. As we have seen, however, the projection is not a theory, but a way of experiencing nature. In order for standardised practices of measurement to develop, there must already exist an experience, a way of understanding nature, according to which such practices make sense. Heidegger attempts to explain the rise of modern mathematical physics in terms of a strengthening existential disposition (*Gemüt*, cf. Heidegger, 1977, p. 19) to experience nature as *a priori* measurable. The question of why measuring practices emerged as a central component of modern science cannot be answered without studying the history of the form of existence in which such practices came to be recognised and accepted as epistemically salient. For Heidegger, then, forms of existence *explain* practices. His existential conception of science thus provides a more fundamental method of explanation than does Ginev's cognitive existentialism.

Ginev is unable to adequately explain what scientific practices are, because he views practice as the bedrock for any explanation. The main conceptual obstacle which prevents Ginev from delivering an adequate account of practice is his errant assimilation of existence to practice. *That* we exist is not, however, a matter of practice. We do not, as it were, practice ourselves into existence. We simply exist. *How* we exist, on the other hand, is very much a matter of practice. Different modes of existence give rise to, and may then be sustained and strengthened by, different configurations of practice. According to Heidegger, the mode of existence most central to modern scientific experience is the mathematical projection which gives rise to specific practices of measurement. These practices, in turn, give rise to specific theoretical models of nature. As I have argued, Ginev's belief that Heidegger problematically identified the mathematical projection with theory, and his attempt to solve this alleged problem by instead identifying projection with practice, fails to acknowledge Heidegger's main phenomenological insights: that both practice and theory can be explained in terms of existence, that the mathematical projection marks a particular form of existence which is prior to both theory and practice, and that this projection is a characteristic feature of modern scientific experience. Ginev's interpretive error is most clearly evinced in his ascription to Heidegger of a theory-driven doctrine of "mathematical essentialism." As I hope to have demonstrated, it would be more accurate to attribute to Heidegger a doctrine of "mathematical existentialism."

5 Conclusion

The historical differentiation of clusters of practice helps to explain how existence gets carved up into different regions of epistemic experience. In the case of science, these experiential regions may be highly sophisticated and specialised, a reflection of the equally sophisticated and specialised configurations of practice which serve to distinguish each unique discipline or subdiscipline. According to my reading of Heidegger, these specialised configurations of practice represent different material articulations of a basic experience of nature as measurable. Heidegger calls this basic experience the mathematical projection of nature. With this, he attempts to put his phenomenological finger on the fundamental disposition motivating modern scientific experience. In his view, existence has a distinct priority over practice, because practices presuppose a mode of existence in which they are experienced as epistemically relevant. Only this explains why scientists working in a particular domain have developed *this* specific cluster of practices rather than some other.

There is no doubt that as scientists begin to articulate and implement a specific cluster of practices, the modes of epistemic existence giving meaning to their work will become correspondingly stronger and more sharply defined. There is, in other words, a positive feedback loop between existence and practice, a loop in which corresponding clusters of practice and modes of existence mutually support and strengthen one another. Through this process of self-stabilisation, a scientific discipline may win its autonomy as a distinct epistemic domain. This is what Heidegger (1962, p. 413) had in mind when he wrote:

> The more appropriately the Being of the entities to be explored is understood under the guidance of an understanding of Being, and the more the totality of entities has been Articulated in its basic attributes as a possible area

of subject-matter for a science, all the more secure will be the perspective for one's methodological inquiry.

This describes a system of epistemic boot-strapping, where an understanding of the being of entities – the existential mode wherein they are experienced as measurable – develops in tandem with an articulation of the basic attributes of those entities – pursued through technical and social regimes of material practice – in order to carve out an increasingly distinct and self-sufficient domain of scientific research.[2] This positive feedback loop of experience and action can only be identified and described if existence and practice are kept analytically separate. Ginev's doctrine of cognitive existentialism blurs the boundaries between the two, and thus loses focus on the object of explanation. Heidegger's existential conception of science is better calibrated to the phenomena, and thus provides a more powerful method of analysis.

Heidegger's existential account also provides a solution to another problem: namely, the problem of *priming*. How does the hermeneutic circle of experience and action get started? Heidegger's answer, on my reading, is that changes in basic patterns of existence give rise to new forms of epistemic experience, and these new forms of experience, together with the desire to explore and strengthen them, motivate the development of new fields of material practice. Because practice presupposes existence, the emergence of new regimes of epistemic practice must finally be explained in terms of changing modes of existence. In my view, such explanations must rely heavily on historical methods, with Edgar Zilsel's historical sociology of science providing one possible example, and Edmund Husserl's phenomenological history of science providing another, quite different, example of how one might attempt to identify the existential origins of science (Zilsel, 1942; Husserl, 1970). In any case, it seems clear that existential phenomenologists of science must work closely together with historians and sociologists of science if they are to achieve a broad and robust account of the existential genesis of science which does justice to its material constitution.

This implication of Heidegger's existential phenomenology of science – that it requires the weakening rather than the strengthening of disciplinary boundaries – may help to explain why the champions of practice-based studies of science have sought so consistently to undermine its credibility by misrepresenting its content. They have fallen in love with practice as a novel explanatory concept, and so lost the critical distance they need in order to properly judge its methodological limitations. Above all, they have forgotten that practices are nothing without practitioners, and so an adequate account of scientific practice must begin with an account of scientists. Accordingly, I have suggested that scientists be studied phenomenologically, in terms of their being-in-the-world, that is, in terms of the specific modes of existence through which they come to make sense of, and so to better understand, the subject matter of their investigations. They do so as practitioners, but they are not nothing without their practices. They first exist as agents who experience their world in interested if initially indeterminate ways, and who subsequently develop sophisticated regimes of practice in order to articulate those experiences in an ever more powerful and determinate fashion. Practices presuppose existence. So long as the champions of practice-based studies of science neglect this fact, their explanatory accounts of science will remain lamentably incomplete.

NOTES

1. Kochan (2011) provides a more detailed discussion of the consequences of this infelicity for Rouse's account of scientific practice, with a specific focus on the issue of realism.
2. Kochan (forthcoming) includes a more detailed discussion of the hermeneutic circle in Heidegger's existential phenomenology of modern science, also with attention to the history of science.

WORKS CITED

Ginev, Dimitri. (2005) "Against the Politics of Postmodern Philosophy of Science." *International Studies in the Philosophy of Science* 19(2), pp. 191–208.

————. (2011) *The Tenets of Cognitive Existentialism*. Athens OH: Ohio University Press.

Hacking, Ian. (1982) "Language, Truth and Reason." In Martin Hollis & Steven Luke (eds.), *Rationality and Relativism*. Oxford: Blackwell, pp. 48–66.;

Hacking, Ian. (1992) "'Style' for Historians and Philosophers." *Studies in History and Philosophy of Science* 23(1), pp. 1–20.

Heidegger, Martin. (1962 [1927]) *Being and Time*, trans. John Jacquard & Edward Robinson. Oxford: Blackwell.

————. (1977 [1954]) "The Question Concerning Technology." In M. Heidegger, *The Question Concerning Technology and Other Essays*, trans. William Lovitt. New York: Harper & Row, pp. 3–35.

————. (1993 [1967]) "Modern Science, Metaphysics, and Mathematics." In M. Heidegger, *Basic Writings*, rev. & exp. edn., ed. David Farrell Krell. New York: Harper Collins, pp. 271–305.

Husserl, Edmund. (1970 [1954]) *The Crisis of the European Sciences and Transcendental Phenomenology: An Introduction to Phenomenological Philosophy*, trans. David Carr. Evanston: Northwestern University Press.

Kochan, Jeff. (2011) "Getting Real with Rouse and Heidegger." *Perspectives on Science: Historical, Philosophical, Sociological* 19(1): 81–115.

————. (forthcoming) "Circles of Scientific Practice: *Regressus*, *Mathēsis*, *Denkstil*." In Dimitri Ginev & Erich Otto Graf (eds.), *Hermeneutic Aspects of the Constitution of Scientific Facts*. Zürich: Collegium Helveticum.

Rouse, Joseph. (1987) *Knowledge and Power: Toward a Political Philosophy of Science*. Ithaca: Cornell University Press.

Zilsel, Edgar. (1942) "The Sociological Roots of Science." *American Journal of Sociology* 47(4), pp. 544–562.

Six

THE UNFINISHED PROJECT OF COGNITIVE EXISTENTIALISM

Dimitri Ginev

1. Chasing the specter of essentialism (A Rejoinder to Professor Arvin Vos)

A

Prof. Arvin Vos criticizes my cognitive-existentialist conception of the status of science's theoretical objects for: (1) a crypto-essentialism in the practical construal of knowledge (manifested, in particular, in my account for theoretical objects by focusing solely on practices); (2) a lack of effective criterion for recognizing anti-essentialism; (3) the insufficiency of the explanation of why theoretical objects are only partially present-at-hand and ready-to-hand in the research process' spaces of representation; (4) a presumed but not explained (trans-contextual) invariance of theoretical objects; (5) a kind of explanatory externalism regarding scientific practices that contradicts the internalist tenets of cognitive existentialism; (6) a neglect of the role which hypotheses play in scientific research (given that very often the theoretical objects come to the fore through the formulation of hypotheses); (7) ignoring the question of whether praxis is prior to theory or vice versa; (8) not addressing the constellation of anthropological (in particular, cognitive-psychological) factors such as intellectual awareness, curiosity, faculty of getting trans-empirical experience which poses necessary conditions of the production of knowledge based on theoretical objects; (9) a reduced account of discovery; and (10) the misunderstanding of understanding through disentangling it from the cognitive experience's procedural manifold, and treating it as a trans-subjective horizon of meaning constitution. Because of these deficiencies, flaws and shortcomings my conception "remains caught in a number of issues that are found in the essentialist framework."

Vos's main assumption runs as follows: There is in the research process a residue of cognitive content that is not dissolvable in scientific practices. The most important ingredient of that residue is the hypothesis. Events like discovery and formulation of hypotheses exceed the scope of practical dealing with what is ready-to-hand in scientific research. (Yet Vos advocates by no means this claim through insisting on the irreducibility of the context of discovery to the context of justification. He rather defenses the autonomy of a "heuristic context of inquiry". The residue I am speaking about is to be addressed in that context.) The aforementioned events do not originate in configurations of scientific practices. Accordingly, the theoretical account of them has to compensate the deficiencies of cognitive existentialism's practical account of science's cognitive structures. In forging such an account, one has to have recourse to science-as-knowledge conceptions. This claim on its part presupposes a strong dividing line between practical and extra-practical part of scientific research. According to Vos, if one fails to address the extra-practical part in terms of science-as-knowledge, essentialism becomes unavoidable. I am not able to see the argument for this claim in Vos's paper. Yet I believe that the dualistic picture of practical and cognitive in scientific research is precisely what one has to overcome in order to achieve a consistent version of anti-essentialism.

B

In trying to systematize my responses to Vos's bunch of critical points, let me start with "the essential anti-essentialism of cognitive existentialism". Imagine an analytical philosopher of science who in trying to elucidate the significance of her/his enterprise offers the following report. The aim of science is the construction of theories which are applied to phenomenal systems within their intended scopes. There are three main purposes of applying theories to such systems. The first of them is the testing of the theory. The second one is the prediction of phenomena that are unknown before the theory's construction. If an experiment designed in accordance with theory's explanatory scenarios succeeds in confirming the validity of the prediction, then there is an independent test of the theory. The third purpose is the explanation of phenomena. (No doubt, the ways of serving these purposes are in relations of mutual reinforcement.) Testing, making predictions, and developing deductive-nomological explanations are normatively constrained procedures. By implication, applying theories to phenomenal systems is unavoidably subjected to a codex of epistemological norms (or, norms of epistemic rationality). It goes without saying that there is a pluralism of alternative (including competing) codices of such norms that very often brings into being inevitable conflicts between different versions of scientific methodology. Nonetheless, behind the plurality of normative methodological systems one is to find a meta-methodological core of non-contextual norms in this method that is eventually determined by what supposedly is the "axiological system of science" (the ultimate values of scientific research).

Prima facie nothing is wrong with this report. It can be found in thousands elaborated varieties in the vast literature of the analytical philosophy of science. Still, there is something wrong. The imaginary report tells us that there is a series of various sorts of determinism taking place (not necessarily in the actual scientific research, but indispensably) in the "context of justification" in which the analytical philosopher undertakes her/his "rational reconstruction". First, by admitting a fixed aim of science (pre-given to any research process) the report postulates a sort of teleological determinism. (Another essentialist assumption in Vos's line of reasoning is that there is a goal of scientific research that can be extracted from the texture of the reading process, and presented as an autonomous aim. This is again a kind of teleological determinism/essentialism. On cognitive existentialism, there are tendencies, orientations and approximations, but not a final goal(s) in scientific research. Since the research process projects its being upon an open horizon of possibilities, it is this process that constantly constructs its own future without assuming a pre-established teleological and axiological order.)

Second, the report assumes that each step in attaining the aim is normatively regulated. This assumption indicates a sort of normative determinism that is vehemently championed especially by those who oppose any naturalization of the philosophy of science. Normative determinism goes hand in hand with a sort of axiological determinism since the coherent advocacy of the former demands a substantiation of the epistemic norms through science's ultimate cognitive values. (The talk of under-determination of scientific behavior by norms, and of norms by values/aims, does not demand a substantial revision of the determinist picture of science. Thus, for instance, Laudan's reticulated model in which this under-determination comes to the fore only manages to create a common framework of all sorts of determinism. Weakening determinism by generalizing the Quinean figure of under-determination does not help one to escape essentialism.) Finally, the report presumes a kind of epistemological determinism since the initial distinction between theories and phenomenal systems is a specification of the subject-object cut as a prerequisite for procedural regimes of getting scientific knowledge. (In this case, it is the foundational representation of the world-out-there that determines the process of cognition.) To be sure, the list of possible sorts of determinism advocated in the traditional philosophy of science is by no means exhausted in the foregoing comment.

But why should one display discontent with determinism of whatever sort? Through each particular sort – so the answer goes – one takes for granted an "essence of science" whose constitution is in need of a philosophical investigation. By skipping such an investigation, one commits the fallacy of hypostatizing cognitive essences. Cases in point for such essences are scientific method, scientific truth, the aim of science, the code of scientific rationality, the meta-methodological codex of scientific research, the invariant logical structure of scientific theory, the structure of explanation, the symmetry principles, and so on. (In the aforementioned formulation I would like to emphasize the term *constitution* which is not to be confused with the *empirical formation* that can be studied in the "context of discovery" or with the *methodical construction* as this gets stressed in constructivist doctrines. (The irreducibility of the *context of constitution* to the traditional context-distinction in the analytical philosophy is of crucial significance for championing the autonomy of the hermeneutic philosophy of science.) I am borrowing the term "constitution" from the phenomenological tradition where it refers first and foremost to the constitution of meaning as a pre-epistemic phenomenon. The production of knowledge takes always place – so a principal phenomenological argument goes – in a world that is already meaningfully constituted. This argument provides at the same time the rationale for distinguishing between constitution and construction. The latter refers only to the production of cognitive structures while the former takes into consideration the role of the meaningful world as a pre-epistemic horizon.) If the philosophy of science resigns from the duty to give accounts of the constitution of what seems to be cognitive essences then it will fail to accomplish the job for which it is entitled – to be both an interpretative theory of what is going on in the cognitive-practical dynamics of scientific research, and a critical theory of self-imposed inauthentic philosophical identities of science's practitioners (in particular, the identities spelled out in terms of metaphysical realism, objectivism and scientism).

To put it in a lapidary way, cognitive existentialism is a family of doctrines that combat all forms of essentialism in the philosophy of science by unveiling the "existential genesis" of each putative and alleged cognitive essence in scientific research. According to a general reproach raised by Vos, my attempt at working out a radically anti-essentialist understanding of knowing falls short since it eliminates significant features about knowing in order to avoid essentialism. (Presumably, these features and events are not stemming out of scientific practices.) This is a paradoxical formulation but it indicates shortcomings of the way I explicate my anti-essentialist position. Yet, I believe, these are not flaws of the very position. So, I am going to oppose Vos's accusation that my approach has failed to break cleanly with the fundamental assumptions about knowing on which essentialism is based. To begin with, let me vindicate my observation that Vos's criticism involves a paradoxical moment. In stressing the aforementioned features and events, Vos argues in favor of the scientific cognition's independence from the epistemic practices in which it is produced. Because of this independence, scientific cognition allegedly brings into being cognitive structures (revealing possibly structures of what is studied) that resist the changes in the configurations of scientific practices, i.e. the intrinsic re-contextualization of scientific inquiry – a claim that is at bottom of all versions of scientific structuralism. At the same time, he purports to be a radical anti-essentialist. The combination is paradoxical since the former claim involves already an essentialist bias: An approach to scientific inquiry that appeals to cognition immune to the ongoing re-contextualization of the process of inquiry is *per definitionem* a kind of cognitive essentialism.

In my rejoinder I will unfold this motif in several directions. In maintaining the existence of science's cognitive content that is immune to contextualization and contextual fore-structuring by practices, one assumes that there is an essence in scientific knowledge on which science's cognitive autonomy rests. This essentialism about scientific knowledge admits that cognitive autonomy consists in a self-sufficient circuit of interrelations between theoretical reasoning and procedural experience. The repudiation of such essentialism involves also a political moment that is in need of further clarification. On the "politics of cognitive existentialism", one has to undo scientism and the other false philosophical identities of science *for the sake of the cognitive autonomy of scientific research*. This tenet seems, as it were, in line with the claim of science's cognitive content that resists contextualization. The advocacy of science's cognitive autonomy is at stake in both doctrines: the conception that the objectivity of scientific knowledge implies the existence of cognitive structures whose validity is beyond any contextual interpretation, and the doctrine of cognitive existentialism. Yet, from the viewpoint of the latter, it is precisely the hypostatization of cognitive structures immune to changes via contextual interpretations that leads to the self-imposition of false (essentialist, objectivist and so on) identities. Scientific communities guided by false identities become easily manipulable, whereby the autonomy of the research process in their domains gets threatened. By combining the struggle against science's false identities with the defense of the research's cognitive autonomy, cognitive existentialism makes critical theorizing an intrinsic dimension of the hermeneutic philosophy of science.

C

Vos makes the case that my denial of essentialism does nothing to explain the origin of science's cognitive structures. I agree. It is the way of scrutinizing the hermeneutic fore-structuring of scientific research from the perspective of the ontico-ontological difference that helps one to give an account of this origin, thereby overcoming essentialism about these structures. In scrutinizing scientific research in such a manner, one is pursuing a much more ambitious task – to integrate the ontico-ontological difference in the actual production of scientific knowledge (or, in science's cognitive body). A genuine hermeneutic phenomenology of science cannot avoid this task. Emil Lensky stresses in the Introduction to this volume the essential similarity between cognitive existentialism's hermeneutic ontology of practices and possibilities and the ontological assumptions of actor-network theory. Yet he ignores (or at least does not comment on) the important dissimilarities between the two approaches. Unlike ANT whose adherents constantly stress various kinds of symmetries and do not make room for introducing the ontico-ontological difference, cognitive existentialism elaborates on a double-hermeneutic methodology in studying practices of scientific inquiry in their potentiality-for-being that is based on this difference.

It is not the denial of essentialism but the way of entering into the hermeneutic circularity of fore-structuring the domains of scientific research that should explain the origin of science's theoretical objects (and more generally, cognitive structures). This observation has much to do with the method of double hermeneutics which I am going later to touch upon.

The nexus of understanding and interpretation (by means of which the ontico-ontological difference is introduced in hermeneutic phenomenology) plays a crucial role in cognitive existentialism. There is a trivial way in which one can explicate this role. Each scientific practice is at once a device of reading something ready-to-hand and a means for objectifying something (as an experimental outcome, a formalized object characterized by the possible values of parameters expressible by measurement scales, an idealized replica of a phenomenal system, a data-model, a physical system's behavior that consists in changing states over time and so on). As a device of reading scientific practice reads something-*as*-something whereby its status of something ready-to-hand gets retained. Here the "as" has the character of a "hermeneutic as". Accordingly, something-as-something in the reading process fore-structures the structure of objectification by means of which something becomes *predicated* as an epistemologically identifiable object. (At this point I am following the hermeneutico-phenomenological thesis that the "interpretative as" fore-structures the "as of predication".) This fore-structuring proves to be (i) understanding of something-as-something within the meaningful whole of a context, tradition, paradigm, intellectual spirit, cultural milieu and so on, and (ii) interpretative articulation of that something-within-the-meaningful-whole.

On this trivial account, the nexus of understanding and interpretation is to be assigned to each particular scientific practice. Furthermore, the nexus concerns the continuity of reading by means of particular practices (reading controlling parameters of experimental systems, measurement scales, instruments, systems of partial differential equations, mathematical codifications of theoretical concepts, etc.). This continuity is contrasted with the discreteness of what becomes objectified. The trivial account, however, does not have resources to explain the contrast itself. The only conclusion one can reach in the framework of this account is that through objectifying something it gets de-contextualized (i.e. isolated from the contexts of reading or disentangled from the ongoing reading process). Finally, in the trivial account, understanding and interpretation are closer to cognitive procedures than to existential phenomena.

Now, there is also a non-trivial way in which one can clarify the role of the understanding-interpretation nexus – a role whose analysis cognitive existentialism attributes to the existential analytic of the constitution of meaning. The non-triviality in this case concerns in the first place the notion of understanding. In the trivial explication, understanding is still a cognitive procedure accompanying all practices of scientific research. It simply means that there is a moment of contextual apprehension of what gets articulated in the domain of research. Yet, there is a hidden connotation in the term of understanding that differs significantly from the meaning of a cognitive procedure. This connotation becomes clear when one says that one understands oneself with regard to the possibilities one can appropriate in one's mode of existence (being-in-the-world). In this case, the whole of the possibilities that can be actualized is the world-horizon in which one exists – the world that is "always already" transcending each and every contextual participation in practices. The actualizations of possibilities lead to the empirical worlds of meaningful realities that might take an infinite number of institutional forms, and remain constantly situated within the world-horizon (i.e. the world of possibilities projected and appropriated by practices). The meaningful realities are not overbuilt or super-structured upon the horizonal circularity of practices and possibilities. They are rather fore-structured by the world-horizon which by transcending the context in which they become constituted characterizes the potentiality-for-being of all modes of existence.

To claim that one understands oneself with regard tothe possibilities one can appropriate amounts to asserting that the meaningful realities one constitutes through appropriating these possibilities are constantly in the making. The empirical being of these realities is entangled in their potentiality-for-being: In their factual (or actual) existence the meaningful realities continue to be projected upon horizons of possibilities. Hence, the empirical being is always already in a state of situated transcendence. (In hermeneutic phenomenology, the expression of "the horizon of possibilities of understanding the world" is tantamount to the claim that "the world as understanding is a projected horizon of possibilities".) At the same time, the appropriation and actualization of possibilities articulates the meaningfully constituted empirical worlds. This articulation has a character of interpretation since it is involved in an ongoing hermeneutic circle – the projection of the meaningful whole and the articulation of units of meaning are mutually presupposing each other. (On this account, the identification of a hermeneutic circle is a criterion for having interpretation. Again, the talk of interpretation here has nothing to do with a particular cognitive procedure. Interpretation is a phenomenon of the totality of being-in-the-world.) In this perspective, the nexus of understanding and interpretation is an ongoing hermeneutic circle that fore-structures each particular meaningful structure (in particular, science's conceptual and theoretical structures) as an outcome of actualized possibilities within-the-world.

D

On a leading motif of Arvin Vos's elaborations, scientific inquiry is carried out by those who are questioning (and creating problem situations). Through scientists' questioning events like insights, formulations of hypotheses and discoveries become possible. (No doubt, these events are hardly to be addressed in the context of justification. They belong rather to Lonergan's heuristics or to what Peirce calls "abductive reasoning".) Only an individual person is able to generate questioning by thinking about something. On a Cartesian argument tacitly employed by Vos, because questioning is a modality of personal thinking, the ultimate subject of scientific inquiry is the individual. (Of course, within Vos's elaborations the latter is taken to be a methodological regulative in focusing on pertinent issues and not the "only real subject" of scientific inquiry. In other words, Vos's position tends to methodological individualism about the heuristic production of scientific knowledge.) On his position, cognitive existentialism's insistence on horizonality, situated transcendence, and trans-subjectivity seems at least quite suspicious. It drives the specter of essentialism from the back door. To be sure, Vos's line of reasoning is intuitively quite reasonable and justified. Yet the intuitive vindication of the suspicion that there is a tacit rehabilitation of essentialism is not a philosophical argument.

The source of the principal disagreement between us is the trans-subjective dimension of scientific inquiry. For me, only by spelling out this dimension, one would be able to overcome essentialism (and a fortiori scientism qua a false philosophical identity of science predicated on various forms of essentialism). I am afraid that there is no room for making sense of trans-subjectivity in Vos's picture of scientific inquiry. In saying this, let me make a small detour to the hermeneutic logic of questioning. The reason for doing this is Vos's accentuation on the role of hypothesis. There are no theoretical objects in science without formulation of hypotheses with testable consequences. And the formulation of hypotheses is not to be disentangled from the manners of scientific questioning. This is why the manners of questioning occupy a central place in what Vos calls (after Lonergan) the "context of inquiry". For Lonergan, questioning determines "the heuristic structure of empirical method". What one supposedly achieves in the context of inquiry is a kind of "erotetic epistemology" that would address the interrogatory aspects of inquiry. Now, the logic of questioning is a topic of prime importance in hermeneutic philosophy as well.

Notoriously, this logic is most profoundly scrutinized from a hermeneutic point of view in Gadamer's *Truth and Method*. The general presumption behind Gadamer's considerations is that the deepest root of questioning as an existential mode is the experience of the situatedness in tradition. What is thus experienced is "meaning that is detached from the person who means it, from an I or a Thou" or from each particular group that experiences its situatedness intersubjectively. (Gadamer, 1989, p. 352) In that perspective, tradition is at once a genuine partner in a dialogue and an experiential reality to which we (as individuals and groups) belong. The experience which effectuates the openness to tradition being so characterized is the experience of the historically effected consciousness. "I" am able to meet the otherness of the other only within that experience: The fusion of horizons between me and the other can take place only in the historically effected consciousness. But then who is the subject of the hermeneutic experience and the historically effected consciousness? The obvious answer is that "we" who belong to (by means of a dialogue with) the tradition. Yet the obvious answer is not the correct one. "We" exemplify or instantiate that subject, but "we" are not identical with it. (What is important for Gadamer is that the particular dialogue between me and you that makes us "we" is mediated by the dialogue with the tradition.) The subject of the historically effected consciousness transcends always already the situatedness of each particular form of "we" in the tradition. No doubt, it bears a resemblance to the Hegelian subject of the total historical experience. However, the conception of the hermeneutic experience implies a completely different (from Hegel's phenomenology) kind of trans-subjectivity (i.e. kind of transcending subjectivity). The trans-subjectivity of hermeneutic experience arises out of the situated transcendence of those who belong to the tradition. Accordingly, the subject of the historically effected consciousness "embraces" both the empirical (individual and collective) subjects who because of their finitude are transcended by the tradition (and who indispensably are situated within it but nonetheless are constantly in a position to change their particular situatedness due to the dialogue with the other) and the trans-subjectivity of the fusion of horizons. The questioning is a "logical characteristics" of this integral subject (or to put it in Gadamer's terms, a characteristic of the "logical structure of that openness which distinguishes the historically effected consciousness"). In the perspective of this openness (or "fore-structure of completion"), every question is to be understood as an answer to a previous question.

To a great extent, the subject of the reading process in scientific research bears resemblance to the subject of the historically effective consciousness. The former is both situated and transcended by the horizon of reading which is handed own through the everydayness of research practices. In other words, a kind of effective-historical situated transcendence operates in the ways in which the implementation of readable technologies constitutes scientific objects. The openness which Gadamer attributes to the historically effected consciousness and hermeneutic experience characterizes also the research process which articulates a scientific domain. Consequently, the questioner in this process is the subject of reading that makes the readable meaningful. Doubtless, particular scientists and scientific teams pose the particular questions of inquiry. Yet the "ethos of questioning" which is inherent to the articulation of a domain is to be assigned to the trans-subjectivity of the reading process – the trans-subjectivity which remains unnoticed in Vos's heuristic conception of inquiry. The questionability of what is under investigation in a scientific domain depends on possibilities projected by practices that can be appropriated within these practices. Since both practices and possibilities are trans-subjective the "ethos of questioning" is irreducible to the heuristic questioning of inventive individuals. I am not trying to underestimate the heuristic questioning. My point is only that the innovative potential of the heuristic questioning (as it is accomplished by creative individuals) is to be understood by taking into account the deviations and transgressions of the established ethos of questioning. Besides, the heuristic questioning is initiated by creative individuals, but it becomes a momentous event in scientific inquiry only when it discloses a new trans-subjective horizon of questioning.

Let me sum up. On cognitive existentialism, the subject of scientific research is the subject of a reading process. It is the reading process where the "logic of question and answer" comes into play. This subject is at once (a) immersed (thrown) in the interrelatedness of practices taking place in the ongoing research and (b) on a par with the very interrelatedness. In this formulation, (b) needs more elaboration. Negatively, it means that the subject of scientific research is not created or determined by configurations of practices already in progress. It would be wrong to say that only practical experience cultivates individuals who in their collectivity (by organizing their activities interactively) become the subject of scientific research. But it would be equally wrong to hold the view that only clever individuals are designing and planning configured scientific practices, thereby becoming the subject in question. The interrelatedness of practices is the ontologically primary reality of existence (the modes of being-in-the-world) to which the subject of the interrelatedness belongs as something on a par with it. In its integrity this subject is to be neither recognized in the individuals nor in the interactively constructed social units. Its recognition requires a reflection that takes into account the ontico-ontological difference. (I will touch upon the status of this subject in my rejoinder to Jeff Kochan as well.)

Here I have to repeat from a different perspective some claims already formulated. On the one hand, the subject is "ontically identifiable" in the choice of possibilities whose ongoing and unceasing actualization envisions the horizonality of scientific research. In this partial identification, the subject of scientific research resides in the subjectivity of cognitive, emotive and volitional characteristics that are indispensable for choosing possibilities. On the other hand, the possibilities that become appropriated and actualized are not to be detached from the subject. Yet these possibilities do not exist as separate units or atoms. They are integrated in open and constantly changing horizons projected by the changing configurations of practices. Because of this projection the existence that is thrown in practices is always already projected upon possibilities. The subject of this existence is "ontologically identifiable" in the trans-subjective horizons. (Accordingly, the notion of trans-subjectivity is to be illuminated in the first place as that horizonality which always already transcends the subjectivity in such a manner that the choice, the appropriation, and the actualization of possibilities become possible.) However, the trans-subjective horizons do not exist *per se*, i.e. independent of the subjective acts of choosing, appropriating and actualizing possibilities. With regard to the hermeneutic circularity already discussed, these horizons only exist through the acts which they render possible.

The whole figure of choosing-possibilities-within-projected-horizons takes on the form of a *hermeneutic circle* between acts that articulate what is ready-to-hand within an interpretatively open world and horizons that transcend contextually performed acts. Entering into this circle requires an ontological reflection (in the sense of ontological difference) on the interpretative articulation. (The talk of the interrelatedness of practices is not to be simplified as a talk about a concatenation, coordination and subordination of particular practices. The interrelatedness of practices is the place where the hermeneutic circle in question operates, thereby constituting meaningfully the primordial reality of being in the world. This is the reality of the constitution of meaning within-the-world – a reality that is not determined by something external to it.)

Thus, the subject of scientific research (as a particular mode of existence) is the circulative unity of subjectivity and trans-subjectivity.

Now, there is an important additional argument that demonstrates why this unity is to be spelled out on the model of interpretative circle. I began my considerations in this section with the thesis that the subject of scientific research is the subject of a reading process. To reiterate, each scientific practice of experimentation, construction and calibration of instruments, development of formalism, statistic interpretation of experimental results, creation of data-models, calculation, semantic interpretation of theoretical models, computer simulation and so on is distinguished by a readable technology that allows reading of data, models, measurements, instruments etc. By implication, each configuration of practices taking place in scientific research is at the same time a "constellation of readable technologies". (Such a constellation is by no means a "superposition" since the interaction of readable technologies creates very often non-linear effects.) Granted that a configuration of practices projects a situational and contextual horizon of possibilities, the readable technologies involved in a constellation are the means of appropriating and actualizing these possibilities. The status of scientific research's subject conceived as a subject of a reading process is best philosophically captured by the figure of situated transcendence. Having a potentiality-for-being (being projected upon possibilities) all entities involved in the reading of scientific research are at once in a situation set up by a configuration of practices (and their readable technologies) and beyond that situation. In working out the figure of situated transcendence with respect to different levels and aspects of scientific research, cognitive existentialism manages to avoid the dilemma between Popper's "epistemology without knowing subject" (the hypostatization of scientific knowledge in its own Platonic world) and the kind of post-Quinean naturalism which reduces philosophy of science to cognitive psychology (the claim that all cognitive events in scientific research have their provenience in the individual mind). Both horns of that dilemma take for granted a relation of determinism. Cognitive existentialism is not a doctrine that seeks for a balance between the two kinds of determinism/essentialism. It does not hold, for instance, that the Popperian third-world-epistemology and the naturalist psychologism must be involved in a mutual compensation of their kinds of one-sidedness whereby the compensation would promote a sort of "feedback determinism". The hermeneutic circle of subjectivity and trans-subjectivity excludes in a radical manner any relation of (causal, teleological or whatever) determinism. This observation presents a further argument of cognitive existentialism against essentialism.

E

To be sure, Prof. Vos is right in criticizing my approach for not offering enough space for treating the real weight of hypotheses in scientific research. Yet he is not right in ascribing to hypotheses independence from and genetic priority over scientific practices. Once Vos has introduced the idea of this independence/priority, he proceeds step-by-step to the full-fledged rehabilitation of the science-as-knowledge paradigm (or, in Vos's terms, the restoration of science's image as a cognitional process). Arguing for this thesis would allow me to pinpoint another basic difference between our positions.

I am not ignoring the philosophical treatment of the structure of scientific knowledge. Yet for me, the dynamics of this knowledge is but the outcome of cognitive practices that are configured with different classes of non-cognitive practices. When Vos attacks cognitive existentialism for its lack of resources to address the issues of scientific hypothesis (and for its inadequate account of what cognition in scientific research is), he tacitly assumes the classical distinction between theory and praxis. However, the interrelatedness of scientific practices is not "the practical" as opposed to "the theoretical". It rather brings to the fore the existential mode of scientific research which becomes articulated in the scientific domains' cognitive structures – the mode of being-in-the-world-by-doing-research. The articulation is due to the "work" done by the hermeneutic circle between subjectivity and trans-subjectivity, which is intimately related to the circle of situated transcendence in the reading process. Cognitive existentialism is not a doctrine that places emphasis upon practices as a strong alternative to cognitive operations and procedures. (The predicate "cognitive" in the title of my doctrine refers to the production of science's *cognitive* specificity within interrelated practices distinguished by readable technologies. As Emil Lensky makes it clear in the Introduction to this volume, cognitive existentialism was originally a doctrine of defending the cognitive specificity of science in terms of a hermeneutic theory of practices.) By implication, cognitive existentialism is also not a particular position within the debate "science-as-praxis versus science-as-knowledge".

On the tenets of cognitive existentialism, the very contradistinction between the theoretical and the practical belongs to the repertoire of essentialist philosophizing. Moreover, one can work out this contradistinction only by assuming a primacy of the epistemic relationship. If one goes on to claim that the practical has a priority over the theoretical, then one hypostatizes in a metaphysical manner a kind of practical essence.[1] This would be a sort of "praxiological essentialism". Cognitive existentialism avoids any reification of practices. The notion of practice does not single out an "independent body" of doings and performances. Practices have a being insomuch as there is an iterative circularity of subjective choices and appropriation of possibilities and the horizons' trans-subjectivity. Practices are not behind or beyond the hermeneutic circularity. They are at once the place where the circularity comes to pass, and the place created by the same circularity. Seen from another angle, practices have a being that becomes revealed by the hermeneutic circularity which articulates the practices' relevant domain.

There is no component of science's cognitive content that resists reconstruction as a derivative of practices of inquiry. All kinds of hypotheses do not make exception. More specifically, the formulation of hypotheses is a *cognitive practice* that has a meaning only in configurations with other practices. (Along with all practices of developing and improving a theory's mathematical formalism, another example of purely cognitive practice is the one of carrying out thought experiments which are more or less reducible either to deductive arguments or to sorts of inductive inference.) Scientists formulate hypotheses if they are provoked by, for instance, some discrepancy or certain discordance of practices of formalization, calculation, measurement, experimentation, construction of instruments and so on involved in a given configuration. (Not to forget that there is a plenty of examples of important research processes that are not cases of hypothesis-guided research.) Vos's considerations are entirely in the (post-empiricist) spirit of the experience-theory correlations. By claiming that the formulation of a hypothesis does not supervene on the accumulation of experience, he follows a central epistemological paradigm of the analytical philosophy of science. The assertion that "discovery leads to the formulation of hypotheses which are composed of theoretical objects" seems to provide the rationale for the privileged status he assigns to hypothesis. (Discovery belongs presumably to a register of events that are not describable in the language of the interrelatedness of practices. But does it indeed belong to such a register? A discovery is only recognizable in a configuration of practices since it is generated by configured practices. On cognitive existentialism, appealing to self-sufficient cognitional process (enclosed in a circuit of procedures) and to knowledge (produced entirely by discrete cognitive acts) amounts to holding a version of the metaphysics of presence.)

Vos goes on to aver that the theoretical objects incorporated in hypotheses and hypothetical explanations are not essences in the sense I reject. I agree. They become essences if the hypotheses in which they are incorporated get treated in the epistemological framework of the experience-theory correlations. Vos advocates their non-essentialist character within a scenario that appeals to statements like (a) the theoretical objects are mental products, and (b) the theoretical objects are experienced as tied to a provisional discovery which will be open to revision or even reversal. In a sense of recapitulation, let me stress that these statements, placed in the Lonergan-Vos's context of inquiry, make the commitment to dubious metaphysical figures of philosophizing – like those of fictionalism and mentalism – unavoidable.

Since the alternative to cognitive existentialism which Prof. Vos adumbrates is essentially influenced by Bernard Lonergan's conception of the heuristic structure of scientific inquiry, a brief comment on this conception would be in place. Lonergan is preoccupied with the circuit of scientific (empirical) inquiry. In contrast to the circuit of mathematical inquiry, it is not immanent. It does not move from initial images through insights to the production of new symbolic images whence higher insights arise. The circuit of scientific inquiry includes action upon "external things",[2] and "moves from observation and experiment to tabulations and graphs, from these to insights and formulations, from formulations to forecasts, from forecasts to operations, in which it obtains fresh evidence either for the confirmation or for the revision of existing views." (Lonergan, 1992, p. 59) Though not immanent, the circuit of scientific inquiry encloses an *order of cognitive procedures*. This order is normatively standardized but in a weak sense. At each stage of performing a class of procedures the possibility of deviating from the standardized order remains open. The weak normative determinism of the scientific inquiry's order of cognitive procedures is associated with the teleological structure of the research process. The deviations from the standardized order are allowed since there are many paths to reaching the final moment of the research process. The weak normativity of the teleological structure implies rather a quasi-algorithmic character of the order of cognitive procedures. It is this quasi-algorithmic way of carrying out scientific research that Lonergan calls "heuristic structure of empirical method", discriminating between classical and statistical version of this structure. ("Heuristic" here is to be understood as a procedural order that is revisable under circumstances. In addition, the structure is heuristic since its weak normativity makes the circuit of scientific research not decidable, in the sense that the norms and rules provide only orientations in moving from one to another set of cognitive procedures. By means of these norms and rules one cannot determine the exact choice of procedure at a certain stage. Presuppositions and expectations are always playing a heuristic role.)

The heuristic structure is at the same time a kind of teleological structure since it is informed by the final stage of the research process, which consists in summarizing all relevant data from a single perspective and achieving symmetry between accurate (sweeping) deductions from laws and exact predictions of new phenomena made in accordance with a theoretical scenario. To sum up, Lonergan offers a prescriptive approach to science-as-knowledge that is akin to (a weak version of) normative epistemology. (Lakatos once suggested that by weakening epistemological normativism, one is practicing epistemological heuristics.) Lonergan's cognitivist approach to scientific inquiry is reinforced by the role he attributes to insight. Not by accident he speaks about the "methodical genesis of insight". The latter is an ingredient of the order of cognitive procedures taking place in scientific inquiry. Insight mediates, for instance, between sets of measurements and the formulation of laws. Following Lonergan's philosophy, Vos depicts a picture of scientific inquiry that is a picture of ordered cognitive procedures (getting initial experience, formulating verifiable and falsifiable hypotheses, experimental testing, replicating experiments, and so on). To stress again, cognitive existentialism differs as a matter of principle from this approach to scientific inquiry because the notion of configured practices is by no means reducible to the notion of ordered cognitive procedures.

Against the background of the foregoing considerations, here is my response to another criticism. According to Vos, my scenario does not offer an explanation of why theoretical objects are only "partially" present-at-hand and ready-to-hand. On the shortest response, they are distinguished by this "dual partiality" just because they always contain non-actualized possibilities. Theoretical objects are always in a state of an *absent presence*. Each contextual actualization reveals a new potentiality, i.e. a "new absence" that in a further context might be made a "readable presence". My account of scientific research is based on the ongoing interplay between de- and re-contextualization of outcomes of scientific practices. Alternatively, a similar account can be given in terms of deferring spaces of representation. The former account is developed predominantly from a viewpoint of a neo-Aristotelian hermeneutics. At stake is the issue of the *phronēsis* of scientific research. The second account seeks for a deconstructionist portrait of science's research dynamics. My contention is that both accounts complement each other.

Cognitive existentialism draws on the doctrine of hermeneutic realism. According to the latter, the reality of being-in-the-world is much more comprehensive than the reality which is out there as something that waits to be represented by mind and conceptualized through the order of cognitive procedures. In this regard, hermeneutic realism is much stronger form of realist philosophizing than traditional scientific realism. Hermeneutic realism is sensitive to the hermeneutic situations of the modes of being-in-the-world. With regard to this sensitivity the methodology of double hermeneutics acquires a central status. Roughly, the double hermeneutics offers two complimentary approaches to the interpretative articulation. On the one hand, it offers a "thick description" of all contingent interpretations taking place in a domain of practices. On the other hand, it looks for the hermeneutic fore-structuring of what gets constituted in that domain. To address this fore-structuring requires to focusing on the characteristic hermeneutic situation of a domain's articulation. Cognitive existentialism forges the concept of such a situation by scrutinizing the forms of reflexivity taking place in scientific research. In a characteristic hermeneutic situation one makes explicit those presuppositions which guide the employment and the harmonization of readable technologies. This endogenous reflexivity aims at discerning the manner in which reading embedded in scientific practices constitutes objects of inquiry. (In my view, Emil Lensky argues correctly in the Introduction to this volume that a characteristic hermeneutic situation becomes manifest via the balance and complementarity of de-contextualizing objectification and contextualizing interpretation taking place in a domain's research process. However, he wrongly tries to reconstruct my concept of such a situation in terms of decision making theory. The "balance" involved in a characteristic situation is not something that might be calculated by means of cost-benefit analysis applied to behavior aiming at achieving two mutually constraining goals. Lensky, an expert in the history of analytic philosophy of science (especially, the post-empiricist tradition), follows actually a trend in post-empiricism preoccupied with the implementation of (heuristic) decision making theory in the epistemological theory dealing with the problematic regarding the justification of methodological normativity. Otherwise, I find Lensky's discussion of the insurmountable differences between the logical-epistemological context of justification and the hermeneutic-phenomenological context of constitution – despite his commitment to the analytical tradition – for a highly illuminative one.)

In reflecting upon the hermeneutic situation of a domain's articulation, those who are involved in the articulation a domain of research attain their consolidation and emancipation qua a scientific community devoted to a certain kind of constituting/reading theoretical objects in established laboratory everydayness. This is a topic that I will discuss in more detail in the coming rejoinders.

2. On the Divergent Being of Science's Theoretical Objects (Rejoinder to Crease)

In this rejoinder I will start right away with a discussion of Professor Crease's conception of formal indication. It is this conception that provides him with a reference system for evaluating doctrines of cognitive existentialism.

A

According to Prof. Robert Crease, perception does not constitute a privileged standpoint in our efforts to approach the status of scientific objects. Perception is always embedded in contextualized practices, and we are able to perceive something "to be real" if we find it to behave in a predictable way within a particular practical context. His version of hermeneutic realism clearly distinguishes phenomenological constitution from all kinds of (social, instrumental-technological, and cognitive) constructionism. At the heart of this version – he dubs it "covariant realism" – is the view that science's theoretical objects (like "electron" or "regulative gene") are invariant under transformation. In fact, this view is a specification of the noetic-noematic correlation in transcendental phenomenology. It is a specification designed to extend phenomenology's constitutional analysis in order to cover the "scientific objects' life". The historical roots of Crease's covariant realism are to be traced – in my view – back to Fritz London's phenomenological approach to the ontology of the objects defined implicitly in the deductive theories. (London, 1923) Kostas Gavroglu (1995) makes the case that many of the ideas elaborated by London in his later treatment of superconductivity and superfluidity as macroscopic quantum phenomena are informed by his early phenomenological wanderings.

Being committed to the noetic-noematic correlation, covariant realism has still much to do with privileging consciousness's intentionality in the constitution of meaningful objects. Husserl's belief that through the method of "free variation" the unfamiliar object will be transformed into a familiar one also plays a prominent role in Crease's view. He appeals also to Cassirer's celebrated conception that the perceptual world always possesses a structure that cannot be reduced to an aggregate of scattered sensations. He has good reasons to invoke this conception in which the "tendency toward invariants" is attributed to the perceptive acts as their principal characteristic. Yet what Cassirer does not address in the course of unfolding his version of holist structuralism is that the structural features of the perceptual world (like all structures formed in human existence) are fore-structured contextually and horizonally. Stressing this fore-structuring does not make the structuralist approach to perception meaningless. It only provides an argument for the priority of the hermeneutic approach in studying the constitution of the perceptual world. Now, with this remark in mind, let me enter into the way in which Crease treats the theoretical objects.

Science's theoretical objects may look differently from different perspectives (or, by being involved in different contexts of doing research). Yet, with regard to each particular theoretical object, the descriptions of "what we see from these perspectives flow together in an orderly way when described by the right set of transformations." (Crease, 2009, p. 227) Making use of covariance as a constraint – in a manner similar to the way in which a theory's basic equations get formulated in a covariant form – allows one to recognize the reality of science's theoretical objects.

Against this background, Crease goes on to rehabilitate early Heidegger's idea of formal indication as a "reckoning with a being-on-the-way". To be sure, this idea belongs to the repertoire of Heidegger's hermeneutics of facticity, and it is originally formulated in a discourse devoted to the pre-theoretical experience. Nonetheless, Crease is convinced that formal indication can be generalized as an approach to scientific entities qua objects that are in a state of being constantly interpreted and anticipated. What one gains by adhering to this approach is the chance to "make explicit the temporality of the grasp" of the objects. (I am not quite sure whether for Crease this temporality is tantamount to the temporality of scientific objects' ongoing constitution.) Conceptualizing the scientific object in a formally indicative manner – so the argument goes – offers "a path of research in advance". The object is never grasped as a pure presence, and hence, it is never totally given. It can be only grasped as having an infinite number of profiles that are potentially to be attained. Scientists hear the language of their discourse as indications allowing them to pick out phenomena which become bodily present in instrumentally mediated ways with possibilities of appearing otherwise given other mediations. (Crease, 2009, p. 228) The investigation of science's theoretical objects comes to the fore as an unceasing anticipation of phenomenal profiles that contextually represent the objects.

Doubtless, Crease's covariant realism based on the formally indicative approach to scientific objects puts into play a promising position in the hermeneutic philosophy of science. However, in the few papers devoted explicitly to formal indication as an approach to science's theoretical objects the position of covariant realism is still underdeveloped, and its depiction suffers from conflating too many issues of different sorts – all of them allegedly playing a role in the attempt at reformulating the Heideggerian concept in terms of philosophy of science. (Without a clear differentiation between the issues which the formally indicative approach handles *en bloc* several undesirable ambiguities may come into play. Crease's confusion of the problematic of the theoretical terms' semantic interpretation with the problematic of theoretical objects' ontological status is a case in point.)

Crease has good reasons for his insistence that making formal indication relevant to scientific research would help one to resolve two aporias in the hermeneutic philosophy of science which cognitive existentialism does not succeed to cope with. The aporias concern respectively the appropriate way of addressing the autonomy of inquiry (more specifically, its irreducibility to changing configurations of scientific practices) and the account of how theoretical objects arise through practices. To be sure, in several respects Crease's approach might suggest better solutions than cognitive existentialism. In spite of this, I should like to express some disagreements with this approach.

It seems to me that formal indication serves the function of a regulative ideal in a sense close to that of German classical idealism. It finalizes "in principle" the research process which, in my view, is always in an open horizon. (I am saying "in principle" because for Crease the research process in each scientific domain is a potentially ceaseless process. Formal indication brings into play a Fichtean regulative ideal whose achievement is always an "unending task".) The formally indicative approach assigns an implicit teleology to the research process that is not congruent with the very idea of hermeneutic openness.

There are potentially infinite ways of contextualizing a theoretical object of science, but each of them is constrained by a (hidden) covariance that ideally regulates (thereby providing a formal indication to) the whole process of the object's contextual interpretations. Formal indication refers to a covariance that is steadily in an upcoming state. In each contextual interpretation of a theoretical object formal indication announces itself as a covariance that is to be expected. Each particular form of covariance might be – *post festum* with regard to the accomplished interpretations – reconstructed as a group of transformations. Yet *ante factum* it is only recognizable as a "feeling of anticipation". This is why having a formal indication manages to temporalize the research process (as being predicated on contextual interpretations of theoretical objects). In the various historical situations the formal indication is characterized by different groups of transformations. But since the formal transitions from one to another group are always possible, formal indication works constantly against incommensurability implied by scientific changes. Crease treats the concept of covariance (and covariant realism) as referring not to the particular groups of transformations but to the ongoing transitions. (If I correctly understand him, formal indication is to be unearthed by reconstructing the diachronic transformations of the groups of synchronic transformations. However, this formulation invokes a structuralist discourse that Crease hardly would accept.)

The regulative ideal of formal indication is a kind of idealization generated by the intrinsic dynamics of scientific research. Thus, the essential value of idealization gets rehabilitated in Crease's conception. Yet this is a quite peculiar rehabilitation. Crease does not regard idealization as an initial trans-empirical codification of possible empirical content that is liable to conceptual-explanatory and experimental articulation. This is why his conception fundamentally differs from the established idealizational conceptions of science. (Thus, the conception worked out by Leszek Nowak and other members of the Poznan school places emphasis upon the idealizing assumptions which supposedly simplify the picture of the investigated reality. Making these [counterfactual] assumptions explicit enables one to formulate procedures of idealizations that determine to a great extent the logic of doing research.)

The regulative ideal cannot be disentangled from scientific practices, and accordingly, stylized as a de-contextualized procedure that provides the absolute target point of scientific investigation. Thus considered, idealization does not contradict the tenets of hermeneutic contextualism. Nonetheless, putting the regulative-constraining ideal first – and thus, admitting that the interpretation of science's theoretical objects hinges on a kind of idealization that is not to be accounted in hermeneutic terms – restricts the scope of the hermeneutic philosophy of science significantly. In focusing on the constraining structure achieved through idealization, covariant realism bears resemblance to structural realism (and other branches of scientific structuralism). Yet unlike the latter, the covariant realist does not confine the constraining structure to the formal-conceptual organization of scientific knowledge, trying then to demonstrate that this structure characterizes reality itself. For Crease, the constraining structure as a regulative ideal takes place in the dynamics of scientific practices. The classical subject-object distinction does not presumably play any role in covariant realism. Thus considered, Crease's conception is by no means something like an epistemological generalization of Carl Gustav Jacobi's canonical transformation theory that aims at "deriving" the theoretical objects' status from studying the transformation properties as they are exhibited within basic physical theories.

Though combatting successfully any kind of hypostatizing idealizations, Crease quite intentionally, I think, keeps intact a residual Platonism in his conception. Indeed, he insists on the contextual existence of science's theoretical objects. Yet he assumes (again in a classical transcendental-epistemological sense) the existence of ideal – de-contextualized and independent of horizon – indicators that operate in the real horizons and contexts, thereby making possible "references to what we encounter". Paradoxically enough, de-contextualized indicators are liable for the contextualization of scientific objects. Yet the price for this "partial restoration of essentialism" has to be paid by making concessions to non-hermeneutic doctrines, and confusing hermeneutic-practical issues with epistemological-semantic problems. Indicative in this regard is again Crease's confusion of theoretical terms with theoretical objects. As already mentioned, he even goes on to substitute the discussion of the objects' status with a discussion of the dual semantics of theoretical terms. By ascribing a kind of semantic parallelism (double meaning of each theoretical term – meaning in a theoretical framework and meaning in the laboratory), he tacitly (and in contradiction with the declared tenets of covariant realism) accepts a traditional epistemological framework.

According to the version of hermeneutic realism I support, science's theoretical objects are entities that exist in particular modes of being-in-the-world. Thus, the constitution of the world meaningfully within these modes has to be at stake in determining the theoretical objects' ontological status. Accordingly, the principled difference between covariant realism and (my version of) hermeneutic realism consists in the difference between the figure of the regulative ideal and the figure of immanent transcendence. It is a difference that involves a contradistinction between the free choices of possibilities in an open horizon and the choices guided by an anticipation of what will come into being that is constrained in a covariant manner. In the latter case, the range of possibilities that can be appropriated is "somehow" always already finalized.

The crux of the matter in Crease's conception is the ambiguity involved in the claim that "the real must not be the function of representation but of our experiences". If under "our experiences" he understands a back-and-forth between meanings embedded in a theoretical framework and meanings constituted in the laboratory, then, I am afraid, covariant realism would land in a Hacking-like interventionism. On the alternative I try to put forward, hermeneutic realism is the view that reality gets always revealed in characteristic hermeneutic situations whereby these situations are inherent to reality. All "experiences" and back-and-forth movements between theoretical and practical meanings get specified by characteristic hermeneutic situations (in the sense in which this notion is construed in my book *The Tenets of Cognitive Existentialism*). Furthermore, I do not agree with the claim that hermeneutic realism is not a constructivism "because the objects that we perceive to be real exhibit invariances that are not under our control". (Crease 2009, p. 224) This could be an argument against constructivism from the scientific structuralism's register, but not from that of hermeneutic philosophy.

In the remainder of this rejoinder I will basically proceed by discussing a historical example whereby my aim will be to flesh out the claims already raised.

B

Imagine a biochemist in 1903 who is engaged with identifying an object called at that time an "enzyme". Let us empathically adopt her standpoint. Being educated (but also indoctrinated) within the established professional background knowledge, "our biochemist" (I will use this expression hereafter) attributes to the term "enzyme" a quite definite content. The neologism "enzyme" had been coined 26 years earlier by Willy Kühne. In fact, he coined this word as a result of a sharp criticism he leveled at the speculations on intracellular ferments. By introducing the neologism, he emphasized the essential difference between "organized ferments" (supposedly living microorganisms) and "unorganized ferments" (enzymes). On Kühne's view, enzymes are agents that can function outside the living cell. The possibility of intracellular enzymes was not ruled out by him. The distinction between ferments (produced by the "living protoplasm") and enzymes (nonliving substances) is still widely accepted in 1903. Though contradicting to a certain extent experimental results obtained in the last decade of the 19th century, the professional ideology of theoretical biology at that time suggests that synthesizing and transporting materials are realized by biochemical reactions within cells but without any catalytic participation of "enzymes" in these reactions. (By "professional ideology" I refer to various suggestions inspired by different theoretical and non-theoretical motives, but not having the character of proven research results. The professional ideology is the system of beliefs of a certain *Denkkollektiv* in Ludwik Fleck's sense.)

For analyzing the status our biochemist ascribes to this theoretical object, we have to take into consideration her historical situation. We have to ask in the first place the question of what she knows about "enzymes" at that time. For sure, she is acquainted with Edward Buchner's discovery that the cell juice extracted from macerated yeast could ferment sugar to alcohol and carbon dioxide[3]. Notoriously, this was the discovery of the soluble enzyme "zymase" responsible for a process of cell-free fermentation. She knows also that there is enough evidence for believing that this enzyme is also an active agent inside the living cell[4]. Our biochemist is well acquainted with the theory of intracellular metabolism proposed by Eduard Pflüger in 1875. Yet this was only a chemical version of the protoplasm theory, and by no means an alternative to it. On the predominant belief (again as part and parcel of the prevailing professional ideology), the attempts at extracting "unorganized ferments" from organs of the animal body that *in vitro* can bring into being the activities of the organs are doomed to failure. Despite this belief, in 1903 our biochemist realizes that the postulates of protoplasm theory are swept out of theoretical biology. (As a consequence, there is a growing conflict between the established professional ideology and new theoretical assumptions and experimental results.)

Our biochemist is looking for a theory that can explain chemically vital processes (digestion, metabolism, respiration, and assimilation) by ascribing each chemical change occurring in the living cell to a specific intracellular enzyme. (Kohler, 1973, p. 185) She is sympathetic to the belief (typified most succinctly by Franz Hofmeister) that all vital reactions are mediated by intracellular enzymes. Our biochemist is enthusiastic not only about Buchner's discovery, but also about some initial experiments giving preliminary evidence that intracellular enzymes carry out oxidation. She has heard – but is still not convinced that the information is reliable – that Gabriel Bertrand had extracted a hydrolytic enzyme that hydrolyzes a substrate to an oxidizable form. She had also taken notice of the discovery of the first enzyme co-factor in 1897.

Quite recently (and almost by accident) she had read in *Berichte der deutschen chemischen Gesellschaft* an article written by Emil Fischer in 1894 in which the first structural explanation of how enzymes work was suggested. Based on her investigation about the reception of Fischer's "key and lock model" (as providing this explanation), our biochemist knows that the model is not valid for all enzymatic reactions. Nonetheless, by employing this model enzymology (and more generally, biochemistry) gets a welcome explanatory independence. The development of the idea that the "geometrical structure of molecules" is responsible for the specificity of enzymatic catalysis encouraged research practices aimed at the elimination of the difference between the chemical activity of living cells and of chemical reagents (in particular, reagents distinguished by molecular asymmetry). (Fruton, 2006, p. 71)

Although our biochemist is still too much indebted to old-fashioned views related to the tradition of Liebig-style "physiological chemistry", her novel views are slowly taking shape due to the outcome of a vigorous public controversy that took place in the years 1897–1902.[5] At issue was the plausibility of the central dogma of protoplasm theory, according to which protoplasm (allegedly identified as a single molecule) carries out all key vital functions. (This controversy called into question not only the premises of protoplasm theory, but the professional ideology as well.) The exponents of protoplasm theory insisted that the active agent in yeast juice was not a specific enzyme but shreds of living protoplasm. Our biochemist continues to hesitate in 1903 as to whether to get rid of the view that (a) enzymes are located outside the cells, and (b) are able to catalyze only hydrolysis. Yet she regards the discovery of zymase as challenging the mainstream of research work in physiological chemistry. Our biochemist is highly interested in the ongoing research work of Sergei Vinogradsky and Martinus Beijerinck on microbial fermentations that provides arguments for dispensing with (a) and (b).

Furthermore, she had read in *Zeitschrift für physiologische Chemie* some reports on enzymes (probably intracellular) that catalyzed oxidations and syntheses. Some colleagues of her go so far to claim that there is a specific intracellular enzyme for each metabolic reaction. She is, however, skeptical about this radical assumption since nobody might verify it. Nonetheless she is enthusiastic about attributing to the enzymes (as the only entities that are the biochemistry's object *sui generis*) more significant physiological functions. She has good professional reasons for that: The success of the "physiological chemistry of enzymes" is the surest way to successful emancipation of biochemistry both as discipline and profession[6]. She is proud of the appearance of the first true biochemical journal (*Biochemisches Zentralblatt*) one year earlier. Like her colleagues she looks at this fact (alongside the formation of Departments for Biological Chemistry at Pasteur Institute and Lister Institute in London) as clear signs of a successful cognitive and social institutionalization of the discipline which is no longer treated as no man's land between chemistry and medicine, or as a trend within physiological chemistry.

In addition to the news related to an intrinsic transformation in the tradition of physiological chemistry our biochemist is literally bombarded with interesting news obtained in neighboring domains – the emancipating from cytology new domain of cell biology in the first place – that soon will prove their significance for various kinds of explorations in enzymology. Thus, she just had gotten the news that mitochondrion is a genuine organelle and not an artifact of technical devices. She follows with a great interest the ongoing work of organic chemists who had made headway by focusing on simpler compounds in studying metabolically important structures like hydroxyl, carboxyl, and amino groups.[7] Our biochemist is surprised that the most vehement support for the existence of chemical agents that catalyze physiological processes outside living cells has come from Emile Duclaux, Emile Roux, and other of the closest disciples of Pasteur. They acknowledge that "Buchner's enzyme" is the first to cause changes in the carbon chains and rearrangements of groups. (This looks most surprisingly in her eyes since Pasteur had been an orthodox adherent to the view that only the living protoplasm may catalyze physiological processes including fermentation.)

Two years ago our biochemist has begun to study Richard Neumeister's attempt at the generalizing of Buchner's idea. Neumeister suggested in 1897 that the fermentation was not effected by a single substance, but by a more complex set of cell constituents. (Fruton, 2006, p. 77) In 1901 Franz Hofmeister presented the initial version of the biochemical enzyme theory of metabolism. The community of our biochemist celebrates Hofmeister's initiative as the first genuine theory in biochemistry in which the concept of enzyme is introduced by definition. Closely related to that theory are Hugo Fischer's arguments (published in 1902) against the old-fashioned distinction between vital and enzyme action. Due to these arguments the view has gained currency in several leading laboratories that enzyme theory should entirely replace protoplasm theory. Yet the accumulation of new results and the resistance of old ideas have created growing crisis in the community's professional ideology. Our biochemist agonizes about the choice between neovitalism (as supported by theories of physiological chemistry that do not make room for the independent activities of particular enzymes) and chemical reductionism.

C

Against the background of this retrospection, I would like tentatively to categorize "enzymes" as species of an *intentional theoretical object*. By definition, an intentional theoretical object is an entity that presumably exists by fulfilling a class of functions that can be fully identified in their empirical details. The object *per se*, however, is never identifiable in its total being. It has always a potentiality-for-being. Only symptoms of its existence become revealed and represented in a situational-contextual-intentional manner. In scientific research such an entity is originally introduced for the sake of filling an "ontic gap" in a domain that is in a process of theoretical and empirical articulation. This entity is constantly (intentionally) given with regard to the intended possibilities of its identification, reading and representation. It is inscribed on a – constantly transcending the research process's situatedness, though in each particular research situation remaining finite – horizon of such possibilities.

Science's intentional theoretical objects have to be distinguished from purely idealized theoretical objects. The latter are introduced solely by means of mathematical idealizations. They do not refer to still "unknown existing entities" whose existence progressively might be specified, but to structures (and objects embedded in those structures) that in their ideality transcend the factuality of what is under empirical study, thereby acquiring a counterfactual status: In order to be empirically valid for all phenomena envisioned, delimited, and circumscribed by a theory (and to offer unique and unambiguous models), the theory's postulates must have a trans-empirical status. This is why the objects these postulates speak about are characterized by the same status. It is the covariance of the theory's formalism (as a mathematical codification of the postulates) that ensures the trans-empirical status.[8] (In other words, the covariance guarantees a kind of morphism between theoretical models and data models. On the standard version of the view of scientific structuralism – in particular, the positions of structural realism and semantic realism – the relation between the two types of models is a relation of isomorphism since only this kind of morphism is a map that preserves the relevant structure as it is revealed by the respective theory. Yet this purely mathematical criterion – in mathematics things are structurally represented by models only up to isomorphism – is too strong for empirical theories [even those in mathematical physics]. This is why several authors [Brading, French, Ladyman, Bueno] have undertaken attempts at relaxing the relation of isomorphism in recent years. The general tendency in these attempts consists in replacing isomorphism with another type of morphism.) Working out mathematical idealizations (and idealized theoretical structures/objects) is a requisite for the introduction of covariance. Thus, for instance, the way in which one relates the metric of space-time continuum to the covariant form of Maxwell equations depends on how one idealizes the behavior of the electromagnetic field.

There is a long tradition (typified most clearly by Hermann Weyl) in theoretical science as well as in philosophy of science of relating covariance and the symmetry principles to the epistemological conditions and criteria of objectivity. As Brading and Castellani (2003, p. 18) argue, there is a simple rationale behind the efforts to connect spacetime symmetries specifically with objectivity: What is objective should not depend upon the particular perspective under which it is taken into consideration, i.e. objective is what remains invariant with respect to the transformation group of reference frames. Peter Kosso provides a more sophisticated argument for the connection between symmetry and objectivity. According to him, every symmetry principle is a principle of objectivity. He states that the argument is "not that values of properties are invariant and hence objective; rather it is relations among the values that are covariant and hence objective." (Kosso, 2003, p. 418) To achieve this objectivity under covariance (as a paraphrase of "invariance under transformation") amounts to disentangling these relations from their contextual interpretations. Symmetry based on covariance is the most effective means of de-contextualizing entities that are mathematically constituted as objects (covariant relations among values) of research. Being thus de-contextualized, the theoretical objects' existence is entirely succumbed to laws written mathematically in covariant form. Yet this de-contextualization is (a) only a particular scientific practice (or a set of such practices, including practices of idealization) that (b) takes indispensably place within a broader contextual configuration of (experimental, formal, measuring, semiotic, instrumental, calculative, etc.) practices – in particular, a configuration of practices aiming at a contextual interpretation of idealized theoretical objects.

It is my contention that Crease's formally indicative approach has to be concretized with regard to the difference between intentional theoretical objects and purely idealized objects. It is worth mentioning that the former may or may not receive a de-contextualization through covariant mathematical interpretation. The intentional theoretical object "enzyme" as discussed so far received historically such an interpretation when covariant mathematical equations of enzymatic kinetics appeared for the first time. The subsequent linearization of these equations by means of various mathematical plots contributed additionally to this de-contextualization. From the viewpoint of the mathematical theory of enzymatic kinetics, "enzymes" acquired also the status of idealized theoretical object. Yet this status hinged heavily on their status as an intentional theoretical object, i.e. as an object that exists through its contextual interpretations within configurations of scientific practices.

4

Let us return now to our biochemist. After 1903 she will become involved (due to her wholehearted devotion to the new domain's articulation) in all of the main configurations of scientific practices in which "enzymes" will be further identified, read off, and represented (through conceptual, mathematical, semiotic, and experimental means) as species of a theoretical object *sui generis*. In a nutshell, our biochemist will participate in practices of material, structural, formal and functional identification of enzymes after the ultimate rejection of protoplasm theory. She will address in 1905 the problem of whether physiological synthesis is due exclusively to enzymes. From a historical perspective, it is this problem that promoted the need of enzymes' fourfold identification. The question we have to ask is whether the respective fourfold research process can be described in terms of Crease's covariant realism.

To begin with, in the first decade of the 20th century our biochemist will become involved in practices of getting a quantitative picture of enzymatic kinetics, i.e. in the search for a formal identification of enzymes. In a full agreement with Crease's conception, the initial formal identification (i.e., identification in terms of quantifiable variables/parameters within a mathematical formalism) had been anticipated by Emil Fischer's work on the intermediate formation of enzyme-substrate complexes. It is the active mass of such a complex that determines the rate of the overall reaction. This observation opened the avenue to formalizable and quantifiable investigation of enzymatic kinetics. In other words, it was Fischer's approach to molecular compatibility within an enzyme-substrate complex that promoted to a great extent the search for mathematical plots of enzymatic kinetics. By the end of the first decade of the last century the main research interest was shifted to a certain extent from the protein structure of enzymes to the kinetics of enzyme-catalyzed reactions. The goal of the emerging search for enzymes' formal identification consisted in establishing predictive relationships between reaction velocities and experimental variables such as enzyme and substrate concentration. At stake was the velocity of these reactions. Attempts at the formation of enzyme rate equations (each of them viewed as a quantitative expression in concentration of reactant or product molecules in time) set the stage for mathematical modeling in enzymology. The most important outcome of these efforts was the Michaelis-Menten formalism which still provides the standard mathematical apparatus in biochemistry.

Enzymes' material identification refers first and foremost to the proteins that can serve catalytic function. In the second decade of the last century there began an intensive inquiry into the various intermediate molecules and enzymes that catalyzed each step in the sequences of possible metabolic pathways. Our biochemist will become familiar with practices of isolating the components of these sequences (enzymes, substrates, cofactors), practices of isolating biochemical intermediates, and practices of synthesizing unstable compounds for use in enzyme assays. During 1908–1914 she will participate Otto Warburg's research team in studying the effect of cyanide and ethyl urethane on the oxygen uptake. This research work led to the material identification of the "respiratory enzyme". At stake was the possibility of deriving the enzymatic activity from the catalytic ability of inorganic lower oxidization. In searching for a material identification, our biochemist will be able to identify oxidases as enzymes that transport molecular oxygen. She will become convinced that the most important problem in enzymes' material identification is that one of how the *in vitro* experiments and solutions can be used in understanding the *in vivo* reality.

Enzymes' structural identification started when Franz Hofmeister enunciated in 1902 the view that the protein molecule is mainly built-up from amino acids. This view opened the avenue to investigating how the differences in amino acid composition of various proteins render them susceptible to degradation by the action of digestive enzymes of the gut. In the eyes of our biochemist the success of this investigation will be the birth of enzymology as a molecular science. By the end of the second decade of the 20th century she will participate in a research team studying enzymes' molecular structures through X-ray crystallography. Later on she will be interested in protein secondary structures. Finally, she will approach the enzymes' geometrical structures that control how the components of biochemical reactions interact. She will be looking for various spaces of representation of how the substrate molecule is complementary in shape to that of an enzyme's active site.[9]

Under "enzymes' functional identification" I mean the precise differentiation of the functions they fulfill. To the initial functional identification of enzymes belongs the search for establishing to what extent the action of an enzyme is affected by the products of the reaction it catalyzes. In studying chemical mechanisms of enzyme-catalyzed reactions, our biochemist will elaborate on experimental practices of exploring an alternative metabolic pathway for a certain reaction, lowering the activation energy, and reducing the reaction entropy. In the 1930s these practices were of prime importance for enzymes' functional identification, especially in the studies of how enzymes fulfill their functions through the energy changes in chemical reactions. Our biochemist will be among the pioneers who will study the regulatory and inhibitory functions of enzymes.

Was our biochemist's research work in 1903 and afterwards guided by "formal indication" in Crease's sense? Yes and no. Yes, insofar as a hidden covariance prompted a rapprochement of trends of inquiry that had been originated in different institutional places and oriented toward different research aims. No, inasmuch as this rapprochement was constantly accompanied by the emergence of research possibilities whose appropriation tolerated and stimulated practices, views, and ideas in each particular trend that were non-transferable and untranslatable in the other trends. Each of the trends of identifying "enzymes" as species of a theoretical object *sui generis* progressively widened its horizon of research possibilities, thereby deconstructing the existing covariance-as-constraint. What formally indicative approach does not take into account is the historical fact that the new possibilities arising out of the appropriation of existing possibilities of the theoretical object's contextual interpretations constantly undo temporarily imposed constraints as related to a certain group of transformations. (I am not going to claim, however, that Crease's conception precludes the opportunity of looking at formal indication as an interpretative process that deconstructs established conceptual, mathematical, experimental, etc. representations of a theoretical object.)

To sum up, there are in the beginning of the last century at least four "formal indications" of how to address the status of "enzymes". Further research in enzymology (after the First World War) additionally pluralizes the formal indications. In the period 1920–1950 there are quite complicated processes of divergence and convergence among these indications. Covariant realism is entitled to give account of the processes of convergence. But what about the processes of divergence (and even discrepancies) among the trends of (material, structural, functional, and formal) interpreting and representing enzymes? I am not able to figure out how covariance might cope with the fact that these trends are distinguished by divergent formal indications. The interpretative-ontological unity becomes threatened, if this intentional theoretical object would have been constituted meaningfully in interpretative contexts and configurations of practices that belong to divergent trends of inquiry.

In enzymology – to continue the discussion of the historical example – there appeared three strategies of overcoming (or at least, dealing with) the destructive divergences in the contextual interpretations of "enzymes'. In fact, these were three types of reaction to the object's interpretative-ontological "demolition" by scattering its existence in an increasing number of interpretative contexts and configurations of scientific practices. These reactions are (i) the "reconceptualization" of enzymology, (ii) the creation of interfield theories which bridge two domains of research, and (iii) the splitting up of domains. All of the three types of reaction took place in the post-war research of enzymes. I will briefly address these strategies/reactions since, in my view, they are not on a par with the formally indicative approach to science's theoretical objects.

The strategy of creating interfield theories in theoretical biology was observed and described for the first time in the mid-1970s. In Darden and Maull (1977), interfield theories function to provide a solution to a characteristic type of theoretical problem of how to construct explanatory models for the relations between the domains. What these authors suggest is an epistemological approach that is to be complemented by a practical-hermeneutic approach to the interfield theories. Cognitive existentialism treats these theories as theoretical practices that transmit in reciprocal manner and recast meaningfully theoretical objects (and other "epistemic things" in Rheinberger's sense) existing in the established domains, thereby constituting intermediary objects and epistemic things. Thus considered, the interfield theories promote new interpretative contexts of reading and representing the intermediary objects. Prominent examples for such theories in the post-war research of enzymes are the thermodynamics of enzyme-catalyzed reactions, the theory of the role of active site flexibility in enzyme catalysis (the theory which transmitted and re-constituted theoretical objects between molecular biology and biochemistry), and the theory of allosteric regulation.

It is the discrepancies between various lines of representing and reading enzymes that make the ultimate identification of enzyme as an object problematic. These discrepancies are the main reason for the under-determination of enzymes' ontic status. The efforts to eliminate such discrepancies provoked the second reaction called by William Bechtel (1984) a "reconceptualization" of enzymology. (On Bechtel's approach, a reconceptualization consists in re-ordering the descriptive and theoretical apparatus used in characterizing the specificity of a domain of scientific research. In a case study Bechtel manages to demonstrate how reconceptualization in the sub-domains of vitamin research and respiratory metabolism facilitated the discovery of the connection between the B vitamins and the respiratory coenzymes.) The splitting up in sub-domains is a strategy that takes place when tenacious (and deteriorating the domain's image) controversies remain unsettled for a long period. In biochemistry this strategy is typically illustrated by the controversy between Peter Mitchell's chemiosmotic theory of the electron transport in respiratory metabolism and the chemical theory about this metabolism.

I am not quite sure whether covariant realism can be extended to cover the strategies of reconceptualization, the creation of interfield theories, and the domain's splitting into sub-domains. The doctrine of covariant realism presupposes a kind of research process's linearity (corresponding to a kind of linearity in the theoretical object's interpretation and representation) that gets violated in these strategies which unavoidably leads to an "ontic pluralization" of the theoretical object's existence. To cut a long story short, the post-war multifarious (and multi-linear) research of enzymes is much better characterized not by appealing to a single formal indication but to a multi-directional appropriation of possibilities.

Finally, let me express my impression that covariant realism is a too optimistic version of hermeneutic philosophy of science. For adjudicating the justifiability of this impression I will first enumerate the necessary conditions which Crease states for establishing the reality of an intentional theoretical object. On the first condition, the real is what scientists can confidently and even unavoidably reach. The second condition refers to the discordance inherent in each present contextual interpretation of a theoretical object. In each context of inquiry the object is "not fully grasped, with hints of another, deeper order just over the horizon." (Crease, 2009, p. 227) According to the third condition, the reality of a theoretical object "cannot be the function of a single theory, or one historical moment". In their unity these conditions have to provide a sufficient ground for a successful identification of the theoretical object's reality. Now, I should like to show that even when the three conditions are fulfilled the theoretical object may prove to be a "non-real entity".

I will again proceed by means of a historical example.[10] The intentional theoretical object I am going to discuss is the "high-energy intermediates of oxidative phosophorylation". In trying to explain what this expression means, I will start with the last component: Oxidative phosphorylation is a highly efficient metabolic pathway of releasing energy in which the mitochondria use their structure, energy, and enzymes to reform adenosine triphosphate (ATP) as the central unit of energy in the cell. ATP is the molecule that supplies energy to metabolism. The energy for the synthesis of ATP is released stepwise via an electron transport chain. A central problem for more than twenty years in biochemistry after the Second World War was the transfer of energy from the electron transport chain to ATP. According to the dominant research program of bioenergetics in the 1950s the energy is transferred from molecule to molecule. The adherents to this program hypostatized the existence of unidentified molecules with high-energy chemical bonds. These high-energy intermediates received at that time the status of a quite important intentional theoretical object. Its identification in the next two decades "would be the central task of the field." (Allchin, 1997, p. 88) Prominent biochemists tried to achieve an interpretation of that theoretical object by means of experimental and instrumental practices of tearing apart the mitochondrion, isolating its components, and throwing them back in a test tube. These practices were subjected to the dogma of "scalar chemistry" – what matters is only the magnitude and not the direction of the reactions.

Within the configuration of scalar chemistry's practices several research possibilities about the object's existence became projected. In the early 1960s possibilities were appropriated that supposedly specified properties of the high-energy intermediates. Theoretical and laboratory practices of making and proving predictions supported the belief in the existence of this theoretical object. For instance, a line of such support was based on tracking radioactive atoms from molecule to molecule through reactions. Another cluster of projected research possibilities were related to the techniques (in particular, by using inhibitors) of forcing certain metabolic pathways to flow through what could be only the intermediates. Allchin convincingly argues that in this period the experimental foundation for believing in the existence of the high-energy intermediates of oxidative phosphorylation was quite rich: One believed that the intermediates exist but because of the complexity of the pathway leading to ATP their reality is not yet identifiable. Leading biochemists announced in the mid-1960s that these intermediates are no longer hypothetical entities. Yet one is to recognize that "all the evidence amassed for a high-energy intermediate molecule was only for concluding that there was an intermediate energy step of some kind." (Allchin, 1997, p. 90) The intermediates had never been found within a certain configuration of practices, though various projected possibilities had been appropriated. They became progressively effaced. Today one is to conclude that the high-energy intermediates of oxidative phosphorylation do not possess reality but they still exist as inscribed in a horizon of possibilities. (The *belief* in their existence is intact in a small dissident group of the scientific community.) Appropriating these possibilities and projecting novel possibilities in new configurations of practices might change the state of affairs with their identification. The incremental disappearance (or better, effacement) of an intentional theoretical object is to be interpreted in terms of cognitive existentialism – an initiative that I will not, of course, undertake here.

Because covariant realism is a too optimistic position regarding the interpretation and identification of science's theoretical objects, I am a little bit skeptical that the approach of formal indication might address properly the problematic of this effacement. In fact, Crease's three necessary conditions for establishing the reality of an intentional theoretical object seem to be fulfilled in the historical existence of the object "the high-energy intermediates of oxidative phosphorylation". For a long period, biochemists believed that these entities can be confidently reached. The research process was constrained by the formal indication of supposed properties and functions of the intermediates. The laboratory inquiry was guided by an anticipation of getting details about these properties and functions. The discordances in the particular contexts of reading and representing the high-energy intermediates became surmounted due to theoretical and experimental corrections in the interpretation that approximated the regulative ideal. In all configurations of scientific practices devoted to the identification of these entities biochemists were able to reason empirically about the specificity of the intermediates. In a full agreement with covariant realism, they managed to show that the intermediates' reality becomes proved by triangulating from numerous experiments. Surprisingly enough, biochemists cultivated their belief that experimental results confirm this reality by advocating otherwise different theoretical positions. Nonetheless, at the end of the day "the high-energy intermediates of oxidative phosphorylation" proved to be (to use Allchin's suitable expression) a "twentieth-century phlogiston".

3. Naturalizing Scientific Practices by Taking into Consideration the Ontological Difference (Rejoinder to Juozelis)

A

The apex of Dr. Evaldas Juozelis's astute critical comments is a series of rhetorical questions. Three of them are indicative of his skepticism towards my philosophical assumptions: Is situated transcendence not a particular manifestation of the primacy of the will? Is there not in cognitive existentialism a hidden appeal to human intentionality that makes its agenda a kind of "humanism"? Does cognitive existentialism have enough resources to avoid the menace of becoming omnipresent meta-science? Adopting the posture of productive misunderstanding of Dr. Juozelis's mode of questioning, I will assume that his rhetorical questions need to be answered. Accordingly, I will put forward the claims that situated transcendence is by no means the asylum which the "primacy of the will" seeks in the time of post-humanism; that in cognitive existentialism human intentionality does not play the role it plays in traditional phenomenology; and that the specter of omniscience poses a much more serious threat to post-humanist naturalism than to cognitive existentialism.

The first claim is easy to defend. Needless to say that Juozelis refers to the will as an epistemological and not a moral notion. In Heidegger's hermeneutic phenomenology, the epistemological aspect of the will is addressed through a revision and extension of a Diltheyan conception. (This is a conception developed in Dilthey's seminal essay "The Origin of Our Belief in the Reality of the External World and Its Justification" [1890].) According to this conception, objective reality is the way of resisting (*Widerständigkeit*) the will when the latter experiences the real. The conscious experience of reality is scrutinized through analyzing the inhibition of the will. This conception has been positively received in hermeneutic phenomenology because it does not presuppose (or postulate) a being of reality as presence-at-hand. Yet Heidegger criticizes it for its lack of resources to offer an "ontological interpretation" of the being of consciousness and the will. In unfolding this critique, he reaches the following conclusion: "The experiencing of resistance – that is, the discovery of what is resistant to one's endeavors – is possible ontologically only by reason of the disclosedness of the world." (Heidegger 1962, 253) This thesis implies that the will might operate and experience the real only within the world. The will and the accompanying experiences (*Erlebnisse*) of resistance do not emerge on their own accord, but take place in the situated transcendence of being-in-the-world. Against this background, one might state that the will can only be manifested under the ontological condition of situated transcendence. I suppose, however, that this is not what Juozelis has in mind when formulating his rhetorical question. In fact, the foregoing considerations reverse the claim implied by this question: The way in which the will operates is a particular manifestation of the primacy of situated transcendence.

The defense of the second claim in terms of cognitive existentialism is a somewhat more complicated task. I would say that hermeneutic phenomenology does not dissolve, but rather absorbs the problematic of intentionality developed by traditional phenomenology. Intentionality becomes integrated in (what Heidegger calls) deliberative circumspection, and there is no need to elaborate on a special concept of it. Admitting that intentionality is a privileged moment in the totality of circumspective dealings with what is ready to hand within the world would entail a revival of the philosophy of consciousness. Notoriously, John Haugeland makes the reformulation of intentionality in terms of hermeneutic phenomenology a central theme of his work. He opposes the "absorption" I just mentioned and accentuates the necessity of retaining the concept of intentionality. Haugeland makes a point of the irreplaceability of this concept when addressing specifically the issue of scientific objectivity in a Heideggerian framework. In this regard, he is greatly inspired by Heidegger's remark that "science has its source in authentic existence."[11] (Heidegger, 1962, p. 415) Authentic existence is characterized by uncompromising resoluteness and radical responsibility.

How these two features are manifested in science as a mode of authentic existence (aiming at the production of objective knowledge) is the question with which Haugeland is especially preoccupied. He admits that a kind of "authentic intentionality" taking place in scientific inquiry makes possible that particular directedness to objects which reveals them "as they are". This intentionality presupposes the capacity for objective knowledge. But the possibility of such knowledge requires the taking of responsibility on the part of the person capable of such objective knowledge. (Haugeland 2013, 264) The responsibility of science – so his argument goes – consists not only in scientists' self-criticism of their individual performances, but in its resources for a critical scrutiny of its practices. It is this scrutiny that is a prerequisite to the possibility of authentic intentionality (as relation between cognitive activities and their objects) and objectivity in scientific knowledge. There is a normativity implied by this prerequisite that is on a par with the normativity of being an authentic and honest scientist. Accordingly, scientists must accept resolutely a kind of critical responsibility. Finally, Haugeland specifies the "authentic responsibility" in science. It concerns the way in which a scientific community revises the laws considered to govern the objects in the domain of inquiry. Authentic responsibility should be taken when the whole domain or discipline is in a state of crisis caused by a revision of fundamental laws. This is a responsibility for the whole, and yet peculiarly personal. The intentionality of someone who does accept authentic responsibility is by definition authentic intentionality.

Thus, the constellation of authentic intentionality, objective knowledge, and acceptance of responsibility is what makes science resolute and responsible, i.e. a mode of authentic existence. On this account, without turning to intentionality, one would not be able to spell out the liaison between objectivity and responsibility in scientific inquiry. This is Haugeland's crucial argument for retaining the concept of intentionality in hermeneutic phenomenology. Yet this argument is, I think, vulnerable. It does not have an exclusive power: One can argue for the claim that science is an authentic mode of existence without rehabilitating the concept of intentionality. Interestingly enough, Haugeland provides (or at least hints at) the rationale for such an option. He (2013, p. 264) writes: "Though I do not agree with Quine that science is 'continuous with' common sense, I do agree that they have the same fundamental character." On my account, this statement ought to be extended/modified as follows: The mode of being in the world by doing scientific inquiry and objectifying the world is not continuous with the existential mode of pre-scientific everydayness, but both of them are predicated on practical-deliberative circumspection. The point of this formulation is that the transition from pre-scientific everydayness to scientific inquiry is a transition not to "objectification without circumspection" but to a new kind of deliberative involvement in practices that requires a resolute acceptance of responsibilities. Once the agents have been made this transition, they are no longer subjected to the anonymous everydayness and the "authority of the They". Despite becoming involved in a new routine (that of Kuhnian normal scientific research), the agents have to take resolutely responsibility for the manner in which they objectify what they are studying. This is why the scenario for the "existential genesis of science" has to be also unfolded as a scenario for a passage from inauthentic to authentic existence. The non-Heideggerian way of coalescing both scenarios is a kernel initiative of cognitive existentialism. And it is an initiative that dispenses with any rehabilitation of intentionality.

Finally, Juozelis voices serious doubts as to whether cognitive existentialism can escape the destiny of becoming a "superfluous omniscience".

The argument he provides in this regard is that the modus operandi of cognitive existentialism (as a program of science studies) amounts to deploying a battery of hermeneutic concepts and figures for re-describing results already obtained by other types of science studies. This argument is to be countered and refuted by demonstrating that there exist meta-scientific issues that are intended by other types of science studies: these issues become clearly visible, formulable, and resolvable in the context of constitution. Roughly, they cannot be resolved without taking into consideration the ontological difference. It should be noted that Juozelis's elaborations give me another opportunity to construe the image of "superfluous omniscience" as well. On Juozelis's account, because of the trivially universal validity of its concepts cognitive existentialism suffers from bad universality. For reasons that will become clear, my contention is that the danger of bad universality threatens much more severely the positions unified under the heading of "posthumanist naturalism".

B

Adopting the stance of productive misunderstanding allowed me to answer some of Juozelis's questions, which are posed not to elicit specific answers but to enhance the skepticism towards my philosophical assumptions. Yet beside the rhetorical questions Juozelis raises the (non-rhetorical) question of what really is the use of hermeneutic phenomenology in cognitive existentialism apart from the formal acknowledgement of (pre-epistemic and transcendental) understanding. In the very formulation of this question, however, he makes use of the expression "human understanding". Thus, he wrongly, in my view, admits that hermeneutic phenomenology addresses understanding anthropologically, i.e. as a human phenomenon. It is this admission that informs his whole exegesis of cognitive existentialism. Actually, the way in which hermeneutic phenomenology ascribes a worldly character to understanding is the point of departure for all post-humanist tendencies taking place in the anti-representational-epistemological philosophizing and the social and cultural studies based on it. (Babich, 2002; 2010, 2014, Crease, 2009, 2012, Theodorou, 2012, Zahnen, 2011)

To prepare some background for my further considerations, I will state that a genuine post-humanism can only be successfully advocated through arguments derived from ontology in which *conditio humana* is not dogmatically hypostatized, but becomes interpretively articulated by means of addressing the question about the meaning of being in the agency's ceaseless constitution of meaningful entities. (With regard to the human agency, this question is to be specified by taking into account the pre-epistemic awareness of being inherent in routine circumspection, whereas concerning the material agency, the question is to be addressed by approaching the co-referentiality of the contextualized utensils. The task of hermeneutic ontology is to unfold systematically the relations between the utensils' co-referentiality and routine circumspection, thereby revealing the unified being of human and material agency.)

Dr. Juozelis displays his trust in Andrew Pickering's posthumanism and Joseph Rouse's naturalism. Accordingly, in his cogent criticisms of cognitive existentialism, he draws heavily on these two doctrines. Juozelis feels perplex as to whether cognitive existentialism should be counted as a humanist or a post-humanist position. I believe he has good reasons for his hesitation. He is completely right in asserting that philosophically cognitive existentialism belongs to neither of these positions. In the final chord of his paper, Juozelis pleads for "a Nietzschean alternative" to the existential-ontological schematization of instrumental rationality. He also claims that the quest for such an alternative is in a harmony with the post-humanist naturalism he subscribes to. Notoriously, the pioneering advocate of such Nietzschean motifs in a post-metaphysical context, using them subsequently as a point of reference for a critique of hermeneutic phenomenology, is Karl Löwith. I am not able to figure out how a Löwith-like philosophy would operate in post-humanist science studies. This is why I will refrain from commenting on Juozelis's concluding suggestion. It is my contention, however, that Nietzsche's conjugation of naturalism and interpretation is an unhappy marriage. Being bound up with perspectivism, this conjugation brings the agenda of interpretive ontology to naught. (Chistoph Cox, [1999, pp. 92–101], whom Juozelis refers to, argues that in order to avoid prominent misinterpretations of Nietzsche, one has to subsume the misleading language of perspective within the richer language of interpretation. However interesting the consequences of this move for Nietzsche's exegesis this move might be, it does not improve the situation in terms of opening an avenue to a genuine interpretive ontology.)

Although it does not share the Nietzschean assumption of the post-humanist theories of practices, cognitive existentialism offers (I dare to say) an original perspective on the post-humanist problematic launched by Pickering's "mangle of practice", Rouse's naturalism, and several other approaches to cultural studies of science, ANT, and STS. In what follows, I am going to discuss those aspects of post-humanism and naturalism which Juozelis implements in his criticisms of cognitive existentialism. The discussion will suggest that the arguments Pickering and Rouse forge for their doctrines demand the introduction of the ontological difference.

C

Joseph Rouse is driven in his research work by the conviction that the fulfilment of the demands of philosophical naturalism is still an impending task. He makes the case that a theory of scientific practices serves the function of coping with this task more successfully than any kind of epistemology. Focusing upon practices, Rouse argues that what matters is not what scientists register through observing nature, but what the phenomena can show them (including the cases in which scientists work hard to create phenomena when they are not able to find them in nature). He carefully decomposes the multilayered structure of phenomena which involves materials, controlled circumstances, signifying elements, detectors for those elements, standardized measures, instruments calibrated to those measures, and various sorts of skillfully performed techniques. (Rouse, 2007, p. 76) The sciences produce reproducible phenomena under laboratory conditions. The constitutive aspect of this production is the causal interactions of all elements being enlisted with the world and within the world. The production of phenomena as arrangements of such interactions helps scientists to articulate the world conceptually.

On this account, scientific theorizing is not a privileged epistemic activity that supposedly organizes and arranges the practices of inquiry. Scientific theorizing is to be understood as a particular practice of modeling various actual or possible circumstances. A philosophical theory of scientific practices is a version of scientific naturalism that aims at ruling out unwarranted philosophical impositions upon science. In producing natural phenomena, scientific practices become part of the world disclosed in scientists' causal interaction with it. Rouse's version of naturalism is in line with the endeavor of early cognitive sociology's to de-sacralize science by rejecting the interpretation of science as an attempt to represent nature from a standpoint outside the empirical world. Rouse (2007, p. 85) argues that we "understand scientific concepts only by understanding the phenomena they articulate. We find ourselves in the midst of the world, and cannot understand it except from within."

I fully accept Rouse's arguments for the claim that the world disclosed by scientific practices has a normative character. At the same time, scientific practices acquire a normative status by being involved in causal inter-actions – or, as he prefers to say in line with his rejection of any standpoint from outside the world, "intra-action" – with natural phenomena. Yet scientific practices are by no means solely the normative part of the world disclosed within the causal intra-action with natural phenomena. Scientific practices are first and foremost the fabric which makes the mode of being-in-the-world-by-objectifying-the-world an existential mode distinguished by a characteristic kind of reflexivity. To be sure, reflexivity is an important issue in Rouse's conception. In particular, tackling this issue helps him to figure out a re-conception of knowing in terms of practices. He also makes the case that the quest for incorporating a political dimension in philosophical naturalism cannot avoid a strategy for coping with reflexivity. Though not fully articulated, the claim emerges in Rouse's writings that a naturalist approach to reflexivity would facilitate the move toward a post-epistemological doctrine of scientific knowledge. Nevertheless, the role reflexivity plays in scientific practices' causal inter-actions with the world remains completely unspecified in Rouse's philosophical naturalism. Of course, one may assume that reflexivity in a naturalist perspective is only a special sort of causal intra-action. But even if one were to proceed in such a manner, one would have to recognize that reflexivity does not operate on the same level as the other sorts of causal intra-action.

Notice that I am speaking about *reflexivity of scientific practices* and not about scientists' (and other agents') reflexivity – a crucial distinction that unfortunately Rouse (and Juozelis) do not take into consideration. Being non-localizable in the empirical network of intra-action, scientific practices' reflexivity appears as that dimension of studying natural phenomena which makes of the network an "emergent" system. As intrinsic relation of practices to themselves, reflexivity is a qualitatively different relation compared with the sorts of intra-action Rouse refers to. Reflexivity cannot be cast in terms of causality.

When particular interrelated scientific practices disclose the world normatively (thereby transforming it into a normative world) they disclose it not *in toto*, but only as a particular domain of doing research. According to a claim that is not at odds with Rouse's naturalism, the interrelatedness of these practices projects the totality of possible causal interactions which circumscribes the domain's scope. The projected totality has no being beyond the ongoing actualization of possible interactions taking place in the particular situations of doing research. The reflexivity of scientific practices consists exactly in the interplay of projection and actualization of possible causal interactions. For reasons I already spelled out in my previous rejoinders, this interplay takes on the form of (ontologically relevant) hermeneutic circularity. A consequent version of philosophical naturalism about scientific practices (and the world of natural phenomena disclosed by these practices) cannot dispense with the hermeneutic circle of reflexivity which always already fore-structures the normative being-in-the-natural-world-of-causal-intra-actions. (In the next rejoinder I will try to show that such a version is absolutely feasible in hermeneutic terms. However radical Rouse's conception, it does not cast doubt on the firmly established view that the hermeneutic approach to science is incompatible with any kind of philosophical naturalism. I will challenge the validity of this view.)

But if interpretive reflexivity – irreducible to causal interaction – is indispensable in scientific inquiry, then one should not try to avoid the need for (a kind of) ontological difference between the level of hermeneutic circularity and the level of causal intra-actions. To stress a tenet of cognitive existentialism, it is a difference that does not imply a Heideggerian dichotomy between the ontic study of the normative-natural world and the ontological study of what is constituted within the hermeneutic circularity. The interpretative fore-structuring of the orders of causal inter-actions is not to be dissociated from the empirical being of these orders. Accordingly, the double hermeneutics which I take to be the appropriate approach to this fore-structuring is not to be disentangled from the empirical study of the causal intra-actions.

Last but not least, there is an essential consequence for the issues of normativity that follows from the foregoing considerations. The point is that the fore-structuring has its own normative force which I call "hermeneutic pre-normativity". (Ginev, 2013c) Ignoring the latter (as a consequence of his failure to reflect upon practices' reflexivity) is a further deficiency in Rouse's normativism.

From Juozelis's Rousean viewpoint, the concepts of cognitive existentialism (as concepts that revolve around the ontological difference) present perhaps a form of unwarranted philosophical imposition upon scientific inquiry. This is not true, but before commenting on such an allegation, I would like to probe the immunity of Rouse's own naturalism from it. Rouse adheres to the view that for the philosophical naturalist, if the established results of scientific research conflict with her philosophical account of science, she must revise this account. Philosophy has no "standing to legislate norms governing science." (Rouse, 2007, p. 64) Rouse discards metaphysical naturalism because it cannot defend through scientific arguments the unavoidable hypostatization of nature. Metaphysical naturalists appeal to different scientific scenarios when they are arguing that mind, communication, action, and the whole of human being are an integral part of nature. But they "cannot just take a plausible story" about the metaphysics of nature from scientific theories. By contrast, Rouse's version of scientific naturalism affirms that philosophy has no independent authority to prescribe how scientific practices should be arranged and executed. The transition from nomological necessity to causal interaction between natural phenomena and scientific practices (or the whole system of intra-action) is the crux of the matter in this version. Only a scientific description of this system of causal intra-action might protect the theory of scientific practices from transgressing the demands of scientific naturalism. Were one to unfold a description in a purely philosophical manner, one would not have a study – continuous with science – of how scientific practices causally produce phenomena.

The time has now come to address the question as to whether the allegedly naturalist picture of the intra-action within the world, which Rouse brings into play is painted exclusively in scientific terms, or to put it alternatively, whether his naturalism does not impose (unjustified) philosophical figures upon scientific practices. I will state right away that there are plenty of apparently non-scientific statements (i.e. non-translatable into scientific languages) in Rouse's scientific naturalism. To which discourse of theoretical science should one attribute, for instance, the following statement: "Science articulates the world, allowing us to tell about it, by developing new patterns of interaction and new ways of talking, together." (Rouse, 2007, p. 78) I am not saying that it is wrong to insist on the priority of a normative conception of what is at stake in science over "more standard appeals to the authority of objects and objective necessity sustained by scientists' and other agents' subjective commitments." (Rouse, 2002, p. 334) But is it consistent with the requirements of scientific naturalism? Again, which are the scientific disciplines capable of providing us with reliable scientific arguments for the statement about the priority of normativity? The very idea that scientific practices are "disciplined patterns" that transform the world so as to articulate it conceptually is hardly continuous with scientific results. Rouse convincingly argues that scientific practices are not activities that scientists impose upon nature from outside. They are rather patterns of causal intra-action within the world. Yet this declaration does not suffice to be a genuine naturalist.

Does he have arguments that the assumptions and the tenets of his philosophical/scientific naturalism are not imposing "philosophical additives" (in Arthur Fine's sense) on science? I cannot find such arguments in Rouse's writings. Nevertheless, he insists that his version attends "more closely to what scientists say and do" than any other version of scientific naturalism.

Rouse has some justified reservations regarding the pertinence of Quine's criterion for "being continuous with science". But he does not suggest an alternative to it. On the criterion I am going to adumbrate, it is not only the actual scientific programs, but perhaps possible scientific developments that might deliver a basis for naturalizing philosophical ideas about science. If this is correct, then to have a philosophy of science that is continuous with science amounts to having a philosophical approach that is capable of entering the leeway of possible forms of scientific inquiry that can become actualized by means of a critique of current research programs. Hermeneutic philosophy of science stipulates further that since scientific inquiry is an interpretive enterprise, a philosophical approach to science that pretends to be continuous with science should allow one to enter in an interpretive manner the interpretive spaces constituted by scientific practices as readable technologies. This double hermeneutics is what cognitive existentialism recommends not only to prevent one from imposing unwarranted constructions on science, but also to eliminate the very viewpoint from "outside". Cognitive existentialism is a radical commitment to the internalist philosophizing about science. It seeks to reveal internal possibilities for doing research, including possibilities that, for various reasons, have been pushed away, sedimented or shifted, after having been projected by scientific practices still influential in the domain of inquiry. Albeit from a slightly different perspective, I will take up the discussion of this issue in the next rejoinder. What should only be stressed for the moment is again the need for ontological difference. It is indispensable when one (for whatever reasons) implements double hermeneutics in studying practices in their potentiality for being.

D

Now, let me turn to Andrew Pickering's version of post-humanism. Generally, Pickering is interested in (what he calls) "posthuman dynamic in the extension of culture". There is in this dynamic a symmetry between the human realm of discursive practices and conceptual structures, on the one hand, and the nonhuman realm of material agency, on the other. Pickering tries to lay bare a process of "interactive stabilization" that characterizes the co-dependence between the "material form of scientific culture" and its conceptual content. It is this interactive stabilization (as involving de- and re-stabilization) that constantly displaces and decenters human agency as a bearer of scientific culture. Given that the strata of this culture are involved in an emergent "mangling", one is not able to assign a foundational significance to any (material, social, or conceptual) stratum of scientific culture. Science is a decentered process of temporally emergent mangling. Scientific knowledge is de-privileged and treated (in a non-representationalist fashion) as something that evolves in "an impure, posthuman dynamics" together with all other material and social strata. All strata reciprocally infect each other.

Scientific culture is depicted as a multiplicity of heterogeneous practices that spans humans (as social actors of different scales involved in networks of relations), cognitive artifacts (like conceptual and representational structures, and elaborated programs of disciplines' reproduction), material artifacts (machines and instruments), and nonhumans (inquired entities and phenomena). "The mangle" operates as interconnected processes of resistance and accommodation that arise in the zones of contact between multiple cultural elements. Pickering (1995, p. 151) blames the strong program in the SSK for missing "the movie in which the social is itself transformed." The social should be exclusively regarded as a particular stratum that constantly interacts with the other strata. He strongly opposes any treatment of the social as an ultimate substance whose illumination supposedly provides firm explanans of what is going on in scientific culture. There is an "emergent dynamics" that goes unrecognized in the strong program. In particular, the explanatory scenarios of this program do not take into account "the mangling of the social".

Yet Pickering's mangle is too extreme in its operating mode, thereby restraining one from introducing distinctions that his approach needs. In a sense, this extremism is vindicated. Without it, Pickering would not have been able to advocate the posthumanist aspects of his approach. Yet the price to be paid for the extreme mangling of what takes place in scientific practice should not be neglected.

Let me be a bit more specific and draw the attention to one of the most important case studies Pickering uses to illustrate how "the mangle of practice" works. I refer to his reading of Hamilton's geometrical interpretation of quaternions. Pickering shows in this study that the constitutive models (consisting in geometrical representation of algebraic structures) are not primitive terms in Hamilton's practice. Their analysis leads to their decomposition into three phases governed by discretionary choices (related to the "intentional structure of human agency") and constrained by machinelike human agency which Pickering calls "disciplinary agency". In this case, one should count as resistance the practice's inability to extend algebra and geometry into three dimensions while maintaining a one-to-one correspondence of elements and operations between them. The modifications undertaken in the phases of modelling epitomize the accommodation of Hamilton's practice. (The invention of a non-routine rule for the multiplication of square roots of -1 exemplifies this accommodation.)

Against this background, the question of what gets mangled is to be answered. On Pickering's (1995, p. 140) suggestion, most "obviously mangled in Hamilton's practice were the modelling vectors that he pursued." The upshot of extensions of conceptual structures in geometry and algebra was the precise structure of the quaternion system. Pickering's case study is interesting because it (in contrast to the other case studies carried out by him) delineates a purely conceptual practice. At issue here is a process of a decentering of the human subject with regard to the disciplinary agency rather than the material agency. The posthumanist aspect of this conceptual practice is to be positioned "between Hamilton as a classical human agent and the disciplines that carried him along." (Pickering, 1995, p. 141) The trajectory of the practice is informed by "the dance" of intertwined human and disciplinary agency. A particular moment of the posthumanist aspect under discussion is that Hamilton's initial goal was radically mangled. For Pickering, this is a highly important observation since it – in line with post-humanism – demonstrates that goals and purposes are not entities that control practices from without. Hamilton's goal is conceivable only from the viewpoint of the disciplinary agency, and more specifically, with regard to the way in which an association between complex algebra and geometry had already been constructed. (Pickering argues that the mangling leads to the effacement of the intentional structure of the human agency.)

To reiterate, Pickering's approach mangles, in my view, a bit too much. But does this approach manage to mangle the ontological difference indeed? On my answer to that question, the mangle of practice is not an actual state, but a tendency in the appropriation of possibilities. It is in a state of situated transcendence. By implication, making sense of the mangle of practice requires a reflection that takes into account the ontological difference.

Pickering convincingly argues that there are phases in the unfoldment of scientific practice in which the practitioners act without discretion. The disciplinary agency holds sway in these phases. In other phases practitioners exercise freedom in choosing possibilities for doing research. This is why strategies of accommodation to resistance may provoke modifications of the disciplinary agency. Yet what is most important in Pickering's approach is that because of "the dialectics of resistance and accommodation" the precise trajectory and the endpoint of a research practice are in no way given in advance. Practice's historical dynamics is not determined. Now, does not this conclusion imply that a research characterized by agency creating alignments between disparate "cultural elements" creates its own regime of temporality as well? Pickering ignores this question. Instead, he directs his attention to the issue of how an interactive stabilization gets achieved through a series of shifts of the initial goal. (The way in which Hamilton extended algebra and geometry has little to do with his initial goal of finding geometrical representations of algebraic structures.) The mangle is described with regard to what becomes mangled. Of course, Pickering is preoccupied with the very process of mangling, i.e. with the dialectics of resistance and accommodation. But he accepts and regards this dialectics at its face value, i.e. only as an actually present series of historical events. By restricting his interest in the interactive stabilization to the factuality of the mangling, he fails to recognize that the dialectics of resistance and accommodation takes place in a much broader "dialectics" of projection and appropriation/actualization of possibilities for doing research by interrelated scientific practices. In order to take into consideration that dialectics, one has to move from the factuality of the mangling to the facticity of the interactive stabilization, i.e. the potentiality for being of the mangle. A crucial consequence of this move would be the change of perspective: At issue will be not the question of how the mangling runs in "the real time", but the (already mentioned) question of how the mangling constitutes its own regime of temporality. This regime as well as the whole potentiality for being of the mangle are only to be addressed through introducing a kind of ontological difference.

Dr. Juozelis argues that cognitive existentialism "in fact does to practice theory what practice theory has been initially committed to the traditionalists." In fact, I do not have any idea about what cognitive existentialism can do to the versions of practice theory Juozelis adopts in his research work. At any rate, it is not a program that seeks to "enrich" Rouse's naturalism and Pickering's posthumanism with phenomenological and/or hermeneutic ideas. To be sure, the cognitive existentialist aims at integrating phenomena like the reflexivity of practices and reflexive re-contextualization in the process of unfolding of practice theory. (Ginev, 2014a) She also intends to demonstrate that a genuine version of posthumanism is to be only achieved by drawing on a kind of ontology that reveals a deeper unity in the intertwining of human and material agency. I do not know, however, whether these initiatives can do something to the established practice theory. The latter is hardly amenable to a reconstruction predicated on a non-dichotomous ontological difference, i.e. the version of this difference advocated by cognitive existentialism.

4. Ways of overcoming the Myth of the Given, or How to Be a Good Naturalist in Post-metaphysical Times (Rejoinder to Gironi)

A

Are really Sellars's alternative to representationalism developed from the viewpoint of his psychological nominalism (all awareness of known entities is a "linguistic affair"), and the hermeneutic reconception of epistemology dramatically opposed to each other? Some followers of Richard Rorty do not think so. I agree with them. Dr. Fabio Gironi is of another opinion. For him, the two strategies of overcoming Cartesian dualism are not to be reconciled. In addition, he admits that Sellars's rationalism might be used as a point of reference for a thorough critique of hermeneutic phenomenology.

Gironi's far-reaching criticisms of cognitive existentialism are targeted primarily on the concept of the hermeneutic fore-structuring of scientific research. He acknowledges that there is a "constitutive fore-structuring", but it is a conceptual-normative rather than a phenomenological-hermeneutic one. Gironi denounces the latter as entailing "impersonal, anonymous conceptual schemes which are immune from revision and correction, and are merely to be passively accepted." (I am not sure whether Gironi wants to present this claim as an implication of cognitive existentialism. At any rate it is completely incompatible with any version of hermeneutic philosophy of science. The ontological figure of situated transcendence leaves no room for pre-given conceptual schemes.) To a certain extent, he plays the old game of opposing contextually transcendental (and normative) epistemology to hermeneutic phenomenology. The game is long forgotten. Yet Gironi is highly inventive and does his best to find an exciting way to continue playing it by changing its rules.

The fore-structuring Gironi pleads for amounts to framing the process of inquiry through a moderately Kantian, normative, epistemology. At stake here is a sort of post-empiricist advocacy of the dependence of practices of scientific research on the pluralism of flexible conceptual-normative frameworks. In line with this engagement, Gironi is convinced that cognitive existentialism trumps epistemic accountability. Accordingly, he wants to rehabilitate and defend the rights of epistemic realism (and epistemological philosophizing) against "Ginev's Heidegger". More specifically, he sees a bright future for normatively designed scientific naturalism. (This sounds like Rouse's story, but goes in effect in an opposite direction. While Rouse treats the conceptual articulation of the world as resulting from practices' causal interaction with the world, Gironi puts conceptualization first.) The harmony between a theory of rational agency, a monist ontology, and a physicalist worldview characterizes further this optimistic vision. (If I understand him correctly this harmonious cooperation should in addition contribute to overcoming the schism between analytical and Continental philosophy as well as to treating the issue of scientism in a proper manner.)

The antidote to Ginev's Heidegger is Gironi's Sellars, an author who unfolds the strategy of pursuing "the difficult but desirable goal of a fusion of the Kantian normative account of thought and scientific naturalism." A few lines further down, I will express my serious doubts about the identity of Gironi's Sellars with the author who once succeeded in reorienting the analytical tradition. But let me for the moment stick to Gironi's author who combines a sort of Kantianism with a sort of scientific naturalism. This is the author entitled to combat "the post-Kantian Heideggerian reversal of epistemological priorities". As a vociferous opponent of "linguistic idealism", Gironi's Sellars offers a reliable standpoint for a realist criticism of cognitive existentialism.

What is at stake in the debate Gironi launches with the gesture of opposing Sellars's conceptual rationalism to Heidegger's ontological pragmatism is, I believe, the issue of how to get rid of "the Myth of the Given". Leading in Gironi's critical considerations is an alleged controversy between those who support "equipmental holism" and those who adhere to the "rule-governed whole of our conceptual arsenal", which is presumably independent of "the practical engagement with our *Umwelt*". In rejecting equipmental holism, he admits that "conceptualization" is the absolute point of reference not only in cognition, but in human life at large. Conceptualization allows human beings to become "sapient reasoners rather than mere sentient organism responding to the environment." The simple question that comes immediately to mind sounds thusly: Does not conceptualization always take place in practices (as integral part of the equipmental world) in which the conceptualizing practitioners are engaged? The affirmative answer, however, still does not suffice to displace conceptualization from the central place it occupies in the views of Gironi's Sellars. Admittedly, conceptualization can serve its grounding function by being embedded in the practices it fore-structures in a normative-epistemic manner. But there is also a more complicated question that the champion of "conceptualization" cannot avoid: Is not "conceptualization" only a heading or a designation for a diversity of conceptual practices (or perhaps more correctly, practices of conceptualization) that do not proceed in isolation but are mingled with non-conceptual practices of various sorts? To answer that question affirmatively is to call into question "the Myth of Conceptualization". (The way in which I tried in my rejoinder to Arvin Vos to deflate and de-privilege the prioritization of hypothesis offers clues to overcoming this myth as a particular form of cognitive essentialism as well.)

Now, the issue I promised to tackle: Is Gironi's Sellars identical with the historical figure of Wilfrid Sellars, the philosopher who inspired the champions of epistemological behaviorism and inferentialism in the post-analytical tradition? Posing this question, I do not intend to cast the correctness of Gironi's interpretation into doubt. I am also not disputing that there is a strong rationale for advocating a Kantian naturalism through re-reading Sellars's critique of representationalism. The question I posed is rather imbued with a hermeneutic sense, as it refers to the "effective-historical reception" of Sellars's work. There is an ongoing "conflict of interpretations" in this reception. In recognizing the significance of this conflict, I should add to the former question the following one: Is not Gironi's Sellars only the upshot of a rather conservative rendering of Wilfrid Sellars's writings? I have the impression that Gironi insists precisely on those aspects of Sellars's philosophy which Rorty and Brandom, in particular, carefully try to remove, or at least to recast in new terms.

The linguistic turn in analytical neopragmatism was inspired by Sellars's thesis that justification is not a matter of a special relation between ideas and objects, but of conversation, of social practice. Once conversation replaces confrontation, not only the notion of mind as Mirror of Nature but also the band between linguistic performances and conceptual frameworks can be displaced. If assertions are justified by society rather than by the character of the inner conceptual frameworks they articulate, then there is no point in attempting to privilege conceptualization as a bastion of epistemologically justified rationality. This is why Rorty's reading of Sellars amounts to advocating the deflation of rationality's epistemological vindication. On Rorty's (1979, p. 174) celebrated formulation, explaining rationality and epistemic authority by reference to what society lets us say, rather than the other way round, is the essence of epistemological behaviorism. The farewell to the myths of empiricism requires no epistemological underpinnings delivered by a neo-Kantian doctrine of conceptualization. Certainly, Rorty would say that it requires no help from hermeneutic ontology either. (That Rorty's way of overcoming empiricism is better served by making use of arguments supplied by hermeneutic ontology is a claim I defend in Ginev [2014b].)

The same de-privileging of conceptualization (at least in Gironi's sense) distinguishes Brandom's influential reading of "Empiricism and the Philosophy of Mind". By insisting that to treat something as "even a candidate for knowledge is at once to talk about its potential role in inference, as premise and conclusion" (Brandom, 1997, p. 123), this reading dispense with the traditional role conceptual frameworks play in knowledge. It seems to me that the arguments in favor of inferentialism are often at odds with the arguments for the primacy of conceptualization. Gironi refers on various occasions in his essay to Sellars's view that the concepts needed to formulate allegedly "foundational facts" turn out to depend on other concepts. I see in this view the strong argument in favor of inferentialism, but am not able to see how it does support (Kantian-naturalist) primacy of conceptualization. If my understanding of Brandom's interpretation is correct, "making a move in the space of reasons" makes the secession of conceptualization as something special in this space redundant. Giving and asking for reasons is not a continuation of a certain initial conceptualization. The horizon of "the practices of using words" is insurmountable. If the talk of knowledge is a talk of (a) which conceptually articulated propositional contents one is committed to, and (b) whether one is in various senses entitled to those commitments, then why should one appeal to conceptualization as a starting-point of developing "methodological linguistic pragmatism"?

The efforts of Brandom and Rorty are oriented towards breaking with static conceptualization, i.e. the one which is detached from inferential justification. I am afraid that Gironi – despite his scrupulous efforts to dismantle the assumption of ultimate frameworks in conceptualization – tries to retain something from the view of conceptual schematism. (At least his commitment to naturalistically recast Kantian transcendentalism implies this.) It seems to me that he is inclined to concede that there are conceptual frameworks preceding (and thus grounding) the inferential articulation. Moreover, this assumption is indispensable if Gironi should like to avoid an infinite regress in conceptualization. I suppose that the constant dissolution of conceptual frameworks would be for him as inacceptable as the uncertainty and instability elicited by "equipmental holism". By implication, there should be conceptual/semantic content that is not inferentially articulated. Maybe I am deadly wrong, but I see in Gironi's interpretation steps towards rehabilitating epistemological foundation not "from the bottom up" (through singling out non-inferential and incorrigible reports about basic facts), but "from top to bottom" (through assuming an ultimate/primary discursive stratum). In this case, the staring-point is from non-inferential (and in this sense, foundational) conceptual frameworks. I can only speculate about Gironi's willingness to admit that there are frameworks that might form an autonomous discursive stratum.

The same argument against the existence of non-inferential beliefs that allegedly serve foundational function is to be used against the assumption of foundational ("from top to bottom") conceptual frameworks: One cannot have one framework that supposedly is the point of departure for conceptualization without having many inferentially interrelated ones. Put alternatively, the foundational use of conceptual frameworks turns out to presuppose inferential uses of concepts. The conservative reading of Sellars tends to look at the concepts as something more than instruments for making inferences whereby this reading paves the way to a hypostatization of the concepts' semantic content. The claim of "the priority of practice over status" which makes Sellars's position a "specific variety of pragmatism" (Brandom, 1997, p. 157) is to be extended to the status of conceptualization as well. I am afraid that Gironi's Sellars is not in line not only with the post-analytical ways of reading "Empiricism and the Philosophy of Mind", but with the Wittgensteinian impetus of Sellars's own philosophizing as well. With Rorty (1979, p. 180) I believe that this philosophizing aims not at an account of how conceptualization operates in human knowledge, but at getting away from the notion of "an account of human knowledge". Yet it is not my desire to subscribe to Rorty's anti-epistemological campaign. With Gironi I claim that the holistic approach to knowledge must not be so radicalized in a behavioristic manner as to display a distrust of the whole epistemological enterprise. The distrust ought to be solely tailored to the epistemological strategies of privileging conceptualization. (I will return to this point later when discussing the relationship between hermeneutics and epistemology in cognitive existentialism.)

Finally, I do not imply that Gironi's Sellars is incommensurable with Rorty's and/or Brandom's Sellars. It is not the case that there are no passages from the former to the latter. But I do believe that Rorty's and/or Brandom's construal entail implications going in a philosophical direction that is not, I suppose, to Gironi's taste. By praising James O'Shea's Sellars-inspired dictum "naturalism with a normative turn", Gironi supplies his line of reasoning with an interesting footnote. He states that his reading of Sellars should result in a philosophical naturalism that takes a sufficiently effective critical distance from "the left-wing or neopragmatist, inferentialist" interpretation of this classic of analytic philosophy. However, there is a price to be paid for any particular interpretation of a twentieth-century groundbreaking thinker. It is a price calculated through the balance of advantages and disadvantages. As regards the advantages, they are clearly exposed in Gironi's essay. The main disadvantage is to be summarized as follows: "Conceptualization" exempted from epistemological behaviorism and inferentialism threatens to rehabilitate the dualism of conceptual scheme and undifferentiated reality. I doubt that Gironi would welcome such rehabilitation. Maybe again I fall short of competence in this matter, but I do not believe that he has come up with any arguments against the aforementioned threat which both of us try to escape.

B

It is not, however, Gironi's reading of Sellars that determines the irreconcilability of our ways to overcoming the Myth of the Given. Much more important in that respect is his parsimonious reading of hermeneutic phenomenology. Whenever he juxtaposes conceptualization that organizes practical experience with "equipmental holism", Gironi does not miss the opportunity to stress the priority of the normatively guided "sapient reasoning" over the "mere sentient adjustment to the environment". He tacitly admits that Sellars's "sentience" (as opposed to "sapience") is tantamount to Heidegger's concernful-circumspective dealing with what is contextually ready-to-hand. Yet both concepts are of quite a different nature, and their parallelizing is by no means a happy initiative. A direct translatability of sentience-sapience opposition into hermeneutic phenomenology's terms is to be ruled out. Deliberative circumspection is seemingly a "continuation" of forms of sentient-animal training, learning, and elementary communication as these forms are approached by ethology. Yet this supposition is incorrect. Sentience that operates on a pre-inferential level is more or less a characteristic of a purely biological (instinctive) behavior. By contrast, circumspective handling within-the-world is a proto-normative, non-biological, comportment. Consequently, deliberative circumspection predicated on the way in which the "hermeneutic as" articulates meaning within-the-world is not a reactive (quasi-instinctive) and a pre-rational kind of behavior. However, this view – as formulated so far – is not sufficiently specified, and because of its ambiguity might be disputed by making use of naturalist arguments. From an objectivist viewpoint – in particular, that of traditional cognitive science – deliberative circumspection is to be completely naturalized and recast in ethological terms. Accordingly, the phenomenological analytic of pre-conceptual behavior can come down to a chapter of ethology. Since Gironi adheres to such a naturalist account, he could make use of it for maintaining the dichotomy between sapient reasoning and sentient adjustment to environment. On this account, pre-conceptual behavior is a manifestation of barely biological life. But even if this were correct, the question remains open as to whether "circumspective sentience" is not a biological behavior that – being integrated in the modes of existence within-the-world – becomes meaningfully (culturally) re-constituted.

Assuming that there is a cultural (and political) re-constitution of human-biological phenomena resonates a motif typically discussed by Giorgio Agamben. In his book *Homo Sacer*, Agamben approaches pre-conceptual behavior – that is outside of the "space of reasons" – as a form of *zōē*, but not a form of *bíos* (i.e. a form of living peculiar to an individual or a group). His elaborations might be useful in making a long story short. Agamben argues that the politicization of bare life as such constitutes the decisive event of modernity. In unfolding this claim, he reflects on some ambiguities of Heidegger's concept of "factical life" as this is guided by deliberative circumspection and distinguished by its fallenness. Agamben is convinced that only when the discussion of this concept is placed in the context of "modern biopolitics" will the concept's true meaning begin to show itself. He accentuates a biological interpretation of factical life which, albeit in conflict with the "hermeneutics of facticity", became quite plausible in Nazi Germany. This interpretation draws significantly on the ontological prioritization of concern and care. By adopting a biological interpretation of the cultural roots of practical life, Nazism became inspired by "the same experience of facticity from which Heidegger departs, and which the philosopher had summarized in his *Rectoral Address* in the formula 'to will or not to will one's own Dasein.'" (Agamben, 1998, p. 152) Yet Heidegger criticizes National Socialism for trying to transform the interpretive nature of factical life into a biological value. This is why Nazism ended up with the incarnation of factical life in an "objective racial determination".

For Heidegger, a consequence from the non-biological interpretation consists in the claim that life (guided by deliberative circumspection) is political in its very facticity. Despite acknowledging the essential difference between both interpretations of factical life, Agamben disputes Heidegger's concept. If one admits that man is not a living being who must transcend himself in order to become political subject – so his argument goes – then one has to concede that "the experience of facticity is equivalent to a radicalization without precedent of the state of exception..." (Agamben, 1998, p. 153) (Gironi insists on the same capability of transcending the sentient-circumspective behavior when he opposes the latter to human sapience: Man has to transcend his "sentient-bare life" in order to become rational subject.) The state of exception is the "bare life of *homo sacer*". On Agamben's account, Heidegger's radicalization of the state of exception is a philosophical legitimization of the "biopolitical turn of modernity" from a position opposite to that of Nazism. Both positions legitimize the biopolitical turn because they advocate the ontological primacy of the indistinction ("indissoluble cohesion") of nature and politics. In modern biopolitics, life is immediately politics. Yet I am not sure that this is a particular flaw of Heidegger's concept. Agamben is interested in preserving the special status of the "bare life of *homo sacer*" in order to develop a comprehensive critique of modern biopolitics. It is this theoretical interest that informs his criticism. (By the same token, Gironi is interested in keeping alive the contrast between circumspective sentience and discursive sapience in order to criticize "equipmental holism".)

In my view, by introducing in an appropriate way the difference between the political and politics, one can retain the thesis of the "indissoluble cohesion" – as it is manifested in deliberative circumspection – without supporting the "biopolitical turn". Yet I have to break the far-reaching story about sentient-bare life at this point. The moral of telling it is that the adherents of hermeneutic phenomenology are – not only theoretically, but also morally and politically –

obliged not only to rule out any attempt at biological (essentialist and deterministic) construal of the circumspective-deliberative, pre-conceptual, being-in-the-world, but also to unveil the specific status of "the political" generated by deliberative circumspection as existential phenomenon. To take up Agamben's motif, one is to state that *bíos* includes not only cognitively organized forms of behavior, but also all forms of meaningfully re-constituted biological behavior as it is otherwise studied by human ethology (and ecology of human populations). Accordingly, pre-conceptual circumspection is not to be conceived as a form of *zōē* that precedes cognition and conceptualization, thus belonging to pre-rational behavior – the behavior which in Gironi's perspective cannot be estimated by epistemological criteria for rationality.

It would be a *contradictio in adjecto* to say that deliberative circumspection is the way in which the constitution of meaning operates in sentient (supposedly biological) human behavior. Any attempt at "naturalizing circumspective sentience" will face this kind of contradiction. (I am not implying that it is not surmountable.) The biological behavior is already transformed within deliberative circumspection. In line with the foregoing considerations, the ontologically primary distinction is not that between sentience and sapience. It is rather the distinction between sentient biological behavior and sentient behavior that (though serving biological functions) is thrown in practices projecting possibilities. In the upcoming discussion, I will follow the claim that deliberative circumspection is the birthplace (and the *Urform*) of reflexivity within-the-world. Reflexivity-as-deliberative-circumspection supervenes on the hermeneutic circularity of the articulation of meaning within-the-world through interpretive appropriation of understanding as world-horizon of possibilities. Needless to say that this is a thesis that cannot be found in *Being and Time*. In a non-Heideggerian manner (that I am not able to unfold here in details) it might be argued that deliberative circumspection is the existentially primary kind of reflexivity that makes possible the interpretive attitude towards the constitution of meaning and the meaningful articulation of the world.[12] Since in the state of deliberative circumspection practices and practitioners are in an "ecstatic unity", reflexivity-as-deliberative-circumspection is a kind in which reflexivity of practices and practitioners' reflexivity are not yet differentiated. Deliberative-circumspective behavior is to be characterized as the "point of departure" of both types of reflexivity. (Furthermore, reflexivity of deliberative-circumspective behavior manifests itself in the duality of this behavior: It is at once embedded in the constitution of meaning and distinguished by the capability to constitute meaning.)

Starting with the classical conceptions of Scheler and Plessner, there is an old belief in the tradition of philosophical anthropology that reflexivity marks the demarcation line between the two kinds of sentient behavior. This is, in my view, a reasonable and warranted belief. It can be spelled out and advocated by appealing to lines of argument developed in various scientific disciplines. Yet the problem with this belief is the idea of demarcation it implies. Hermeneutic philosophy (like many other theoretical discourses) is against drawing a firm demarcation line between the two kinds of sentient behavior. Though only human sentient comportment is doomed to be meaningfully constituted and to constitute meaning, thus having a world-horizon, each attempt at elaborating theoretical terms with empirical content for explicating this demarcation fails. There is really a play of differences (and not a basic dichotomous difference) that differentiates between meaning-constituting sentient comportment and teleonomic-functional (biologically sentient) behavior. In refraining from a dichotomous formulation, I would suggest a moderate way of differentiating: Sentient behavior is typically manifested by non-human species and non-typically (mediated by constituted meaning) by humans. Human sentient behavior guided by deliberative circumspection is always embedded in practices and possibilities, thereby embodying meaning (possibilities projected and actualized by practices). In this way of differentiating, reflexivity retains its status as a distinctive feature of sentient behavior thrown in practices. This feature gets manifested, however, through the aforementioned play of differences.

Until now at issue was the complexity of circumspective sentience as irreducible to pre-rational behavior. There is also another phenomenological-hermeneutic scenario for addressing this complexity. According to it, deliberative circumspection occupies an in-between position. In mediating between horizons of understanding that are "always already" there and the procedures of knowledge production, deliberative circumspection "transmits" the trans-subjective interpretative appropriation of understanding (as wordly horizon) to the realm of human cognition. Deliberative circumspection is "immersed" in the space between the trans-subjective horizons of cognition and procedural cognition, but can never be localized as something that is to be objectified as empirically identifiable presence-at-hand. (Before acquiring epistemic status, understanding and interpretation are primarily trans-subjective phenomena of meaning constitution within-the-world.)

In uniting both scenarios discussed so far, one might tentatively state that – from the viewpoint of hermeneutic phenomenology – deliberative circumspection is (a) the existentially primary kind of reflexivity (and "source" of all kind of reflexivity in human cognition), (b) the existential medium of*phronēsis* rationality, and (c) the originary place of the political. Let me briefly comment on (b) and (c).

In *Being and Time*, circumspection is the guide for practical concern in the "everyday-average" mode of existence. Here again I have to underscore that Heidegger ascribes a constitutive role of practical concern only to the pre-theoretical mode of being-in-the-world. For him, scientific inquiry is predicated on the primacy of "theoretical seeing", and there is a room for practical concern only to the extent to which it is derivable from this seeing (as synonymous of epistemic representation). By contrast, the starting-point of cognitive existentialism is the non-derivability of practical concern from cognitive faculties (including the faculty of conceptualization) even in the intellectually most sophisticated forms of social behavior like scientific inquiry. A basic goal of cognitive existentialism is to develop a hermeneutic theory of deliberative circumspection as it operates within the scope of practical concern in scientific inquiry.

To return to Heidegger, circumspection as the guide for practical concern correlates with "deliberating". On the phenomenological account suggested by him, deliberating is the way of enacting ("bringing closer") entities that are ready to hand by interpreting them circumspectively. It is the nexus of circumspection (as opening a horizon of cognition) and contextual deliberating, and not "mere representation", that affords (and fore-structures) knowledge through conceptualization. Deliberative circumspection is characterized by its own (pre-epistemic) rationality whose discussion looms large in the development of hermeneutic philosophy. On Gadamer's "urbanization" of Heidegger's ontology, the rationality of deliberative circumspection coincides, by and large, with *phronēsis* (as non-derivable from *epistēmē* and *technê*). Accordingly, this is the reflexive rationality of the originary way of transcending the sentient-instinctive-reactive adjustment to environments. Those who become doomed to deliberative circumspection (and constitution of meaning) have no more natural environment (e.g. a specific ecological niche). Instead, they are horizonally open to the world, which is in a constant state of meaningful articulation.

From the viewpoint of a (hermeneutically oriented) social theory, the rationality of deliberative circumspection reveals itself as the sphere of practical intersubjectivity which differs from the sphere of intersubjectivity governed by politically organized power. (If, and to the extent to which, the former is independent of the latter, there is a leeway for initiatives not controlled by politically formalized mechanisms. With regard to the difference between the political and politics – the so-called "political difference" – the sphere of practical intersubjectivity is where the political becomes generated, while the other sphere produces the formalized mechanisms of conventional politics.) Practical concern guided by deliberative circumspection (and *phronēsis* rationality) acquires the special status of the "originary place of the political" in hermeneutic phenomenology. Precisely because deliberative circumspection is distinguished by pre-epistemic reflexivity, the subjects of practical concern are able in their routine everydayness to "hear the call of conscience authentically", i.e. to free themselves from the anonymous authority of average public opinion (*das Man*), and to become in a resolute manner morally and politically – in the sense of "the political" – responsible individuals. (In line with the discussion of Haugeland's view in the preceding rejoinder, the move to authentic mode of existence is a prerequisite for scientific inquiry.) Not in complete accordance with Heidegger's scenario, I would say that the existentially primordial reflexivity involved in deliberative circumspection takes the public idle talk away and allows one in reticence to hear the call for a resolute responsibility in one's being-in-the-world.

On hermeneutic phenomenology, deliberative circumspection is intimately related to discursive articulation. At this point, I have to address briefly the use of linguistic arguments in our debate. Gironi cogently unveils the potential of Sellars's philosophy to dismiss "the belief in any allegedly immediate, self-authorizing and epistemically foundational inner states." In this regard, he underscores the "epistemic priority of linguistic performances within rational communities". However, Gironi is not an epistemological behaviorist and he is not willing to compromise the autonomy of normative epistemology, reducing it to a descriptive analysis of linguistic behavior. Then, how the epistemic priority of language-in-use is to be advocated? Obviously, all linguistic performances (including all communicative interactions) take place in a linguistic medium that has more horizonal rather than substantial character. In Gironi's Sellarsian idiom, this medium is called "socially regulated intersubjective language". It is a prerequisite for thought episodes and enjoys logical primacy in "the order of knowing". But if so, then the epistemic priority of linguistic performances hinges crucially on the ontological priority of linguistic medium. If cognition were to depend on language-in-use (not in the determinist sense of Sapir-Whorf linguistic relativity), then epistemic subjects (and their faculty of conceptualization) would have become emancipated within intersubjective language.

I am quite aware that Gironi would not go on to accept these conclusions. But do not they follow from his line of argument? It is my contention that by supplying his intersubjective anti-Cartesianism with ontological assumptions about language-in-use, Gironi would serve better his intention to defend the autonomy of (linguistically designed) epistemology without making concession to foundationalism. (No doubt, the diagnosis that the epistemological critique of foundational representationalism should be complemented by arguments from hermeneutic ontology of language [*Sprachhermeneutik*] is to be extended to Sellars's psychological nominalism as well. The "linguistic affairs" that make knowledge possible are not purely mental events. They transcend mentality and have a being of their own.)

C

Gironi's central accusation against cognitive existentialism is the "rejection of objective reality". There is a rather simple (almost trivial) response to that accusation: Cognitive existentialism does not reject objective reality, but only claims that this reality is to be seen from the viewpoint of (scientific) practices rather than to be postulated as the cause of all practical being-in-the-world. Making sense of objective reality from this viewpoint (and not from God's eye view, for instance) is the main thesis of hermeneutic realism as a counterpart of cognitive existentialism. Hermeneutic realism is not an amended version of internal realism since it disavows from the very outset Putnam's distinction between externalist and internalist perspective. This is a distinction that uncritically presupposes the primacy of epistemology. While according to internal realism, one must first develop an epistemology capable of recognizing that human beings are functioning as part of reality and that it is impossible for them to just step outside, thus becoming pure observers of the whole reality, hermeneutic realism argues that human beings become epistemic subjects only by being involved in the reality of practices. By implication, hermeneutic realism requires (for its accomplishment as a programmatic doctrine) not a special epistemology, but an ontology of practices' potentiality-for-being. (This is not an anti-epistemological statement. It implies that only the situated subject, who is able to produce embedded and embodied knowledge can play the role of an active epistemic subject. "Active" means, in particular, being able to intervene in what is under investigation.)

On hermeneutic realism, the phenomenon of "having access to objective reality" should be thought over in a manner that has again much to do with reflexivity of practices. In the upcoming considerations, the already mentioned distinction between this kind of reflexivity and the researchers' reflexivity will come to the fore. The way in which the practitioners-as-epistemic-subjects perceive and apprehend objective reality supervenes upon reflexivity generated by the practices in which they are involved. More specifically, if the phenomenon under investigation becomes circumscribed by two conflicting (or at least incompatible) configurations of experimental, instrumental, and conceptual practices, then it will manifest itself as different families of objects.[13] The reflexivity of practices shows itself in the way in which it contributes to the settling of the clash between conflicting/competing ways of reading, representing, and "saving" the phenomenon. The clash is settled by delineating a new context of approaching the phenomenon, i.e. by re-contextualizing the process of inquiry. It is a re-contextualization implied by the intrinsic "grammar" of practices.

By "grammar" of practices I mean the tendency of creating alignments (configurations) of practices within their interrelatedness. This tendency is due to practices' endogenous reflexivity. The concept of "grammar of scientific practice" is discussed in Chang (2011). Chang's main assumption is that scientific inquiry consists of actions (physical, mental, and discursive operations). He tries to approach this concept from the viewpoint of Wittgenstein's later philosophy. Accordingly, scientific practice is analyzed in terms of epistemic activities, each of which is defined as a coherent set of actions. A merit that deserves mentioning is Chang's treatment of scientific thinking as consisting of activities as well. Thus, there is no residue in scientific inquiry that resists decomposition in activities and eventually in actions. Yet, at the end of the day, Chang's grammar offers rather a new version of operationalism. It is a version that extends Bridgman's doctrine by replacing the view of science as a private activity with a view that science's epistemic activities are founded on second-person relationships. The philosophical grammar of scientific practice comes into being thanks to these relationships. This grammar is concerned with the rules of communicating discrete units (actions) to particular recipients. Put differently, Chang's project pays exclusive attention to what can be operationalized as elements of scientific practice. Such a grammar, however, fails to do justice to the horizonal integrity of scientific inquiry. The sum of discrete actions and activities does not exhaust the entirety and the meaning of science-as-practice. If one leaves the territory of operationalism, one has to look for a "grammar" that takes into account that continuity of actions and activities which precedes the differentiation of discrete units. Only such a concept of grammar could be pertinent to reflexivity of scientific practices. Actually, this concept is closer to what is called "hermeneutic logic" in the Diltheyan tradition than to Wittgensteinian grammar.

Practices in their "grammatical interrelatedness" are inherently reflexive since they have the trans-subjective capability of re-arranging their configurations that contextualize the process of inquiry. Generally, the re-contextualization (and the new contextual interpretation) of what becomes circumscribed in practices is the outcome of practices' capability of reflexivity. One can only "see" the reality "behind" the diversifying objects by coming to grips with this kind of reflexivity. How the intrinsic grammar of scientific practices invokes a re-contextualization (of what is investigated) is an issue that the practitioners of normal science tackle on a regular basis. In so doing, they specify the validities of the phenomenon's contextual interpretations. Objective reality always becomes revealed through contextualizing practices, and (to reiterate) the very contextualization is an implication of practices' endogenous reflexivity.

Doubtless, reflexivity of practices constantly provokes researchers' reflexivity. For instance, a clash between scientific practices may call into question established categorial assumptions, conceptual frameworks, schemes of formalization, tacit knowledge, etc., thereby leading to researchers' new understanding of objectivity. The most extreme case of calling established visions of objectivity in question comes to the fore when the same configuration of scientific practices may constitute alternative objects (say, waves and particles) in the process of studying a phenomenon. Consequently, reflexivity of contextualizing practices becomes more intensively entangled with reflexivity of researchers. The interplay of the two forms of reflexivity enables one to look for a complementarity between alternative contextual interpretations obtained within the same configuration of practices. The ways of perceiving, apprehending, accessing, and coping with objective reality cannot skip this interplay. As I mentioned in the preceding section, deliberative circumspection is the "point of departure" of both types of reflexivity. Yet this circumspection is not "neutrally tuned and disposed" to practitioners' reflexivity. As "practical understanding" (in a sense closer to that of Bourdieu) deliberative circumspection is the milieu in which the practitioners' reflexive attitude becomes articulated.

Two issues are of great significance in this regard: first, the issue of when and where researchers' reflexivity should take place, and second, the issue of how far this reflexivity should proceed. It is a form of reflexivity that problematizes the tenability of the researcher's "prejudices" (in a Gadamerian sense) about the objectivity of what is seen as objective reality. In this regard, it might play a rather destructive role. As a rule, the researchers of natural phenomena avoid such reflexivity. Yet it becomes indispensable when reflexivity of practices of inquiry requires a definition of objectivity vis-à-vis an ongoing diversification of contextual interpretations. Actually, these considerations offer another way of doing justice to the concept of characteristic hermeneutic situation. The interplay of reflexivity of scientific practices and researchers' reflexivity – the interplay through which standards and criteria for objectivity in investigating natural phenomena become formulated and employed – brings into being a balance between objectification and contextual interpretations in the research process. By means of this balance the practitioners involved in the research process are able to fore-have, fore-see, and fore-grasp the objects they constitute through readable technologies. This stable tendency of fore-structuring in the articulation of a domain of inquiry is what I call a characteristic hermeneutic situation.

Against this background, the attention has to be drawn again to the relevance of the ontological difference to the issue of how practitioners make sense of objective reality from the viewpoint of the practices in which they are involved. The interplay between reflexivity of practices and practitioners' reflexivity is on a par with the constitution of the primary meaning of the natural phenomena under investigation (including the meaning of the whole domain of these phenomena). This constitution takes place in the continuous stream of changing configurations of interrelated practices. On a corollary, meaningful entities become constituted in scientific inquiry through the continuous re-contextualization of what is ready to hand in these configurations. To say that the entities thus brought into being never become present at hand in the stream of practices is to say that their meaning is inter-contextualized, i.e. scattered over all possible (interconnected) contexts in which the entities are manipulated/read/textualized. The continuity of the stream of re-contextualization (corresponding to the changing configurations of practices) is to be contrasted with the discreteness of what becomes thematically isolated (procedurally de-contextualized) as objects of inquiry. Yet the procedural de-contextualization – or what Heidegger calls "objectifying thematization" – is always achieved within a configuration of practices. Thus, paradoxically enough, all de-contextualization in scientific inquiry is necessarily contextualized. (Phrased differently, de-contextualization in the process of inquiry is only a modality of re-contextualization.)

The domain of discrete (de-contextualized) objects is the ontic domain of inquiry delineated in a characteristic hermeneutic situation. (A domain's entities become discrete objects by exempting them from the stream of re-contextualization, i.e. by depriving these entities of their inter-contextualized meaning. This exempting/depriving is another name of objectification. To reiterate, only being thus exempted/deprived the objects acquire the status of being present at hand.) If one goes on to put the question of how such a domain becomes disclosed, one would have to approach the tendency of fore-structuring in the continuity of practices of inquiry. This reflection upon the characteristic hermeneutic situation is an ontological reflection, irreducible to any kind of ontic reflection. In practicing reflexivity that problematizes the tenability of their "prejudices", the researchers are actually engaging themselves in such an ontological reflection. This is why the latter is by no means a privilege of the hermeneutic philosopher of science. The researchers are ineluctably reflecting upon the characteristic hermeneutic situation of the research process, when in a reflexive manner they control the balance between contextualized interpretation and objectification in order to "confront" objective reality. From the viewpoint of being-involved-in-practices-of-scientific-inquiry, there is no way of making sense of objective reality but via the ontological difference.

D

I take Gironi's reproach that cognitive existentialism is at odds with epistemological theory of scientific inquiry quite seriously. His objections that my doctrine makes no room for approaching cognition epistemologically need to be met separately. To that effect, I will extend the view of de-contextualization, adapting it to cover aspects of the process of inquiry-as-textualizaing. Accordingly, de-contextualization will refer not to particular objects of inquiry, but to whole "texts" constituted by readable technologies. To begin with, Gironi puts forward the claim that cognitive existentialism "tends to reject the necessity of epistemology". My entire work can be characterized as a quest for a complementary unity and (given the primacy of the ontological difference) an optimal division of labor between hermeneutics and epistemology. Cognitive existentialism is a post-epistemological but by no means an anti-epistemological doctrine. Admittedly, the complementary unity I mentioned is a topic with too many ramifications in order to be discussed on this occasion. Let me pick out only one aspect of the whole discussion which illustrates the importance of non-representationalist (holist) epistemology to cognitive existentialism. To a certain extent, this is an aspect that concerns the synergy of epistemological structuralism and interpretive post-structuralism in my approach as well. (Ginev, 2013b)

Cognitive existentialism depicts scientific inquiry as a process of textualizing. The claim is raised that textualizing (reading entities that are ready to hand in configurations of practices-as-readable-technologies) is ontologically prior to the particular texts. (In reversing the text-reading relation and putting the reading that textualizes first, cognitive existentialism adopts some of the principles of Jauss's reception theory.) Texts are embodiments of actualized possibilities within the continuity of projection and appropriation of possibilities. Accordingly, they are at once situated within and transcended by the practices which project and appropriate possibilities. In their being-situated and being-transcended simultaneously, the texts circulate across conceptual, experimental, instrumental, computational, and other kinds of spaces of representation. (A whole text is not to be found in a particular space, but in an integral circle of deferring spaces of representation that corresponds to a configuration of readable technologies.) In this circulation, the texts constantly change their meaning. The constant being-in-a-situated-transcendence implies that the texts are to be only relatively delineated as discrete units. The delineation is due to a temporary "extraction" from the continuity of textualizing. In admitting that there are relatively completed circles of deferring and interpenetrating spaces of representation (each of them corresponding to a configuration of practices), one might identify a particular text as what becomes meaningfully constituted through the readable technologies operating in such a circle. (To reiterate, a text is what becomes totally represented in an integral circle of deferring representations of results obtained by the readable technologies operating in that circle.)

Thus, for instance, a text is the total representation of an unstable subatomic particle in the circle of its (a) experimental reading through a particle accelerator, (b) conceptual representation in terms of quantum field theory, (c) representing through data of measurements of mass, mean lifetime, decay mode, spin, etc., and (d) computational space determined by the Lagrangian describing a certain interaction. This circle/configuration, in which what becomes represented in the space of a readable technology is already deferred to the space of a next technology, is only relatively completed since at each moment it might trigger off a new circle of spaces of representation and configured readable technologies through which a new text will be brought into being. More specifically, if practices of experimentation are that endpoint which reaches the starting-point of the circle, its completion would trigger a new instrumental design and a new reading of experiments, thus eliciting a new circle of textualizing.

Yet scientific inquiry would be impossible without periodically "drawing out" the texts from their circulation in configured readable technologies. By disentangling the texts from the deferring spaces of representation (and changing configurations of readable technologies), one manages to bestow the texts (as discrete units) on stable semantic content. The way in which they become disentangled is the way in which the ongoing constitution of hermeneutic meaning in scientific inquiry becomes transformed into semantic content of isolated outcomes. But as discrete and disentangled bodies, the texts are not only the semantic units of the structuration of a domain's cognitive content. The structuration of semantic meaning inevitably takes place in (and is conditioned by) the ongoing constitution of hermeneutic meaning. In this regard, the difference between semantic and hermeneutic meaning parallels the ontic-ontological difference. On this account, texts are also epistemic models of reality revealed by scientific inquiry as a specified mode of being-in-the-world. It is the versions of the so-called "scientific structuralism" that suggest scenarios of how epistemic modeling to be related to the theories' semantic interpretation.[14]

To return to the example I barely mentioned, the text obtained by reading/representing an unstable subatomic particle in a configuration of readable technologies is an epistemic model of the reality of, say, strong nuclear force. In contrast to scientific as well as to structural realism, I would not go on to claim that this model represents particular entities or the structure of reality. It rather presents a structure that remains invariant with regard to a group of formal transformations. It is a structure that is (ontologically) disclosed by the interrelatedness of practices which characterizes a mode of being-in-the-world, and a structure rendered (ontically) meaningful through the totality of its semantic models. More specifically, this structure can be semantically interpreted through data models of measurements. Thus, the epistemic relation between (what is accepted as) the theoretical and the empirical is mediated by this semantic interpretation. Notice that the text contains both the structure and its semantic interpretation. This concept of text does not require for its semantic interpretation the postulation of extra-textual entities. Since texts are the epistemic models, there is no epistemic relationship extrinsic to texts. (Without trying to enter into sophisticated debates in the analytical tradition of scientific structuralism, I would only like to point out that my preference is for the view that the objects of inquiry [as de-contextualized and "extracted" from the stream of practices] exist within the structures presented by the epistemic models. This preference is closer to structural realism than to traditional scientific realism. On the ontic version of structural realism, the fundamental physical objects are relata of a physical structure conceived of as a network of concrete physical relations. These objects do not have any existence independently of the structure they are part of. [Esfeld and Lam, 2011, p. 143] The champions of this version provide us with good arguments that quantum field theory, general relativity, and other theories in physics support the ontic priority of structure over the objects involved in it as relata. However, one should not forget that this priority is not absolute since it presupposes the ontological disclosedness of the domain in which the structure can be interpreted semantically.)

Against the background of the foregoing considerations, the profile of holist epistemology launched and vindicated by cognitive existentialism can be concretized. These considerations give in the first place, I believe, a clue regarding the way in which cognitive existentialism approaches the complementary unity of hermeneutics and epistemology in the process of textualizing. Concretizing the profile of holist epistemology is to be accomplished in terms of the doctrine of textualizing. The epistemic models are discrete bodies isolated from the continuity of meaning constitution. They present the reality (of a domain) which is already disclosed meaningfully within a mode of being-in-the-world. Regardless of how effectively the epistemic models have been detached from the ongoing process of textualizing, they remain always already fore-structured by the continuous projection and appropriation of possibilities within the stream of changing configuration of practices. The epistemic modeling mediated by the theories' semantic interpretation forecloses the introduction of basic epistemological dualisms that hypostatize the subject-object cut. The (non-foundational) epistemic oppositions – objectifying-objectified, signifying-signified, representing-represented, etc. – promoted by this modeling are always contextualized. One only makes sense of such oppositions by taking into consideration the continuous interplay between textualizing and isolating particular texts.

E

I would like to conclude this rejoinder with a note about the perspective of "naturalizing the hermeneutic philosophy of science". This note takes into account my critical comments on Juozelis-Rouse's kind of naturalism I suggested in the previous rejoinder. There is a widespread conviction that philosophical hermeneutics and (any form of) naturalism are irreconcilable enemies. This conviction is well epitomized in Dr. Gironi's essay. In stating that "the avoidance of the mystifying subject-world dualism can take a naturalist rather than a hermeneutic slant without subscribing to empiricist dogmas", he makes no secret of his belief that the philosophizing which revolves around the ontological difference and "equipmental holism" is the exact antipode to the programs looking for naturalizing philosophy of science. No doubt, there are a lot of versions of philosophical naturalism that can do successfully the job of overcoming Cartesian dualism. Rouse's naturalism is a superb case in point. If a kind of naturalism can accomplish the philosophical work – the destruction of the traditional metaphysical assumptions in doing philosophy of science – for which the hermeneutic theory is supposedly entitled, then the latter loses, as it were, the legitimation of its existence. Yet an option – despite its striking absurdity – is still intact: Hermeneutic theory can beat naturalism on its own territory – the territory of naturalizing philosophy of science. This option implies something like a "hermeneutic naturalism".

It goes without saying that the locution "hermeneutic naturalism" sounds like an oxymoron. However, a program for a "naturalizing hermeneutic philosophy of science" (and thus making it an empirical enterprise) seems to me a completely conceivable and feasible task. Its outcome would be a kind of hermeneutic naturalism. Here are some arguments in this regard. (To a great extent, the view of hermeneutic naturalism as outlined here is the outcome of my reading of Azarya Polikarov's philosophy. As a student of Polikarov I was deeply impressed by his attempt to make branches of philosophy of science continuous with non-algorithmic [heuristic] programs of AI.[15])

On the conception of "naturalizing epistemology", the study of the nature of knowledge ought to be made continuous with science. Quine's naturalized epistemology is broadly speaking an unavoidable consequence of his critique of logical empiricism. If there is no purely analytical meta-knowledge entitled to provide (logical and normative) justification of first-order empirical knowledge, then the study of the cognitive processes of knowledge production and knowledge evaluation ought to be relegated to the empirical inquiry of cognition, i.e. to psychology. Hence, epistemology is but a "chapter of psychology" as a part and parcel of natural science. On the epistemology naturalized in terms of psychology, it would be better to discover how science is developed and learned than to fabricate a fictitious rational reconstruction to a similar effect. Quine's demand "to be continuous with science" involves several of his conceptions (in particular, behaviorism, and indeterminacy of translation) which are dubious from various perspectives. Yet in one respect the demand raises no objection: To look for a philosophy of science that is continuous with science amounts to transforming this philosophy into empirical studies. Indeed, empiricizing hermeneutic philosophy of science (by means of double hermeneutics) would lead to a quite different concept of empirical basis as compared with the concept congruent with the objectivist standards of naturalizing (analytical) philosophy of science. Nonetheless, it would result in a kind of hermeneutic naturalism.

Hilary Kornblith (2007, p. 40) is absolutely right when arguing that by making epistemology continuous with science, one brings with this initiative "a reconceptualization not only of the subject matter and methods of epistemology, but of the epistemic status of epistemological theories as well." Against the background of this observation, I would like to state that there is a significant asymmetry between naturalizing epistemology (cum epistemologically designed philosophy of science) and naturalizing hermeneutic philosophy of science. The former changes the status (and perhaps the structure) of epistemological theories, but leaves the status of the empirical (both as empirical basis of these theories and empirical data of the first-order theories) untouched. By contrast, naturalizing hermeneutic philosophy changes principally the status of the empirical. Roughly, it recasts factuality – which in post-empiricist perspective consists in the theoretically laden (experimental) facts – in facticity, i.e. "facts of existence" that are meaningfully constituted within practices and entangled with horizons of possibilities. While factuality is a present actuality (or actual presence) that from a semantic point of view might be designed as semantic models of a theory, facticity never ceases to be predicated on a potentiality-for-being and never becomes a pure presence. *The construction of scientifically relevant facticity is the task of naturalizing hermeneutic philosophy of science.*

Hermeneutic-phenomenological theorizing about scientific research can be only continuous with scientific theories that integrate in its structure a kind of constitutional analysis of meaning. The place where theories pertinent to "naturalizing the context of constitution" (without reducing it to the context of discovery) should be sought is evidently (the phenomenologically oriented) cognitive science. From their inception (in the mid of the 1980s) these theories are constructing their empirical basis by means of phenomenologically devised methods that take into consideration the contextual constitution of meaning in the intelligent behavior.

The unfolding of a coherent historiographical narrative about the development of these theories is a task of prime importance for those who champion philosophical hermeneutics. This narrative should involve several stages (related to paradigm-shifts) in the development of research programs in the so-called "Heideggerian Artificial Intelligence". In the perspective of naturalizing hermeneutic philosophy of science, these are stages of preparing a phenomenologically shaped empirical basis of cognitive science. Let me mention the main of them. In 1986 Winograd and Flores published the book *Understanding Computers and Cognition* which, in my view, was the birthplace of the research tradition preoccupied with approaching artificial intelligence from the viewpoint of hermeneutic phenomenology. Following Gadamer's dictum that prejudices are biases of human beings' openness to the world, Winograd and Flores tried to develop a view of practical (deliberative) understanding – as at once opposed and intimately related to discursively articulated understanding – that might become instrumental in programs looking for alternatives to the mainstream trend in artificial intelligence. They insisted that programs in AI must get rid of the symbolic paradigm's assumptions. Cognition does not rest on the manipulation of symbolic representations that are to be regarded as referring to items of the world out there. Winograd and Flores (1986, p. 71) conceived cognition not in terms of an "activity in some mental realm" but in terms of practices rooted in "the pre-reflective experience of being thrown in a situation in which we are always already acting." The investigation of these authors presents the first stage in the delineation of empirical basis of studying intelligent behavior that is entangled with constitution of meaning. The next stage took place in Rodney Brooks's approach to modeling intelligent behavior that copes appropriately in a timely fashion with changes in its dynamic environment. Brooks (1999, p. 90) reached the conclusion that "there need be no explicit representation of either the world or the intentions of the system to generate intelligent behavior." It is not the inner processing, but "the state of the world" that determines the complexity of behavior of a system. Philip Agre, whose investigation presents typically the next stage, took up the theme of designing engineering constructions that could emulate human behavior predicated on embodied practices. He was preoccupied with the question of how to integrate indexicality in AI research. In this regard, Agre (1997, pp. 243–246) proposed the creation of a non-objective type of ontology, which he called "deictic ontology". The latter is basically a description of what becomes constituted by "deictic intentionality", and can be defined only in indexcal and functional terms. On Agre's account, deictic ontology informs a new kind of computational research that is strongly alternative to the mentalist computational theory of representation developed in the symbolic AI. Agre's program alluded to the non-localizability of cognition to mental processes. "The kinds of things" which can produce knowledge are (indexical) units of interaction between environments and beings endowed with cognitive abilities. A further stage of empirical research in the "Heidegerian Artificial Intelligence" is presented by Michael Wheeler's combination of action-oriented representations that cope with the locally arising un-readiness-to-hand with "causally spread" distributed mechanisms of cognitive behavior.

This combination allows one to translate the Heideggerian concept of "smooth coping" within the world (Wheeler's locution for Heidegger's average mode of being-in-the-world) in AI's language. I fully agree with Hubert Dreyfus (2007) that only a progressive research program (in Lakatos's sense) of Heideggerian AI can be counted as a successful alternative to the traditional and connectionist programs in cognitive science.

The construction of empirical basis in this fashion does not amount to developing theoretically laden facts that can be experimentally verified or falsified. Yet this construction is still not tantamount to developing facticity as empirical basis pertinent to hermeneutic-phenomenological (scientific) theorizing. The job of attaining such a basis is only to be accomplished through naturalizing hermeneutic-phenomenological philosophy of science – a job that demands the integration of scientifically unveiled facticity in the doubly hermeneutic study of scientific inquiry. Only via this naturalizing the facticity of discursive-practical being-in-the-world disclosed within the phenomenologically oriented theories of cognitive science can be further elaborated as basis for studying empirically the "existential phenomena" which hermeneutic philosophy of science delineates. This is the very idea of hermeneutic naturalism.

5. The battle for mathematical existentialism and the War of the Heideggerian Succession (Rejoinder to Kochan)

Dr. Jeff Kochan is one of the few cognitive sociologists and social historians of science who try to build bridges between science studies and hermeneutic phenomenology. I admire his work. It is my deep conviction that the ontological trend in STS can profit significantly from the way in which he introduces phenomenological ideas in empirical programs. It remains to be seen whether he will be able to combat successfully the symptomatic tendency (in recent science studies) toward sociologizing existential ontology.

A

Dr. Kochan accuses me of committing an unnuanced reading of Heidegger's existential conception of science. I am afraid that Kochan's critique has little to do with my analysis (in Ginev, 2011) of the diverse readings of the existential conception of science in *Being and Time*. But before countering this critique I would like to stress my disagreements with his way of adjusting existential analytic to the standards of a doctrine he dubs "minimal realism". In so doing, I have to repeat arguments I developed elsewhere. (Ginev, 2012) Kochan believes that the realist assumptions of Heidegger's hermeneutics of facticity can be adequately reconstructed by divorcing the claim that nature has a natural-kind structure from the semantic claim that science's theoretical terms have factual references. For him, the very procedure of divorcing "minimizes" realism in a manner that approximates Heidegger's realist assumptions. Heidegger (to be sure) does not explicitly maintain a claim about natural kinds, but he argues tentatively – so Kochan's argument goes – that the plausibility of nature's natural-kind structure is amenable to an essentialist and an existentialist interpretation. Heidegger's putative minimal realism is informed – if my reading of Kochan is correct – by the leeway of possibilities for coping with "the real" (*das Wirkliche*) as this leeway is determined by the two (metaphysical) interpretations of nature's reality supposedly distinguished by natural-kind structure.

Be that as it may, the move from scientific realism to minimal realism (as a move promising a new exegesis of Heidegger's position) invites the following objection: However minimized is the metaphysical claim of realism, it remains part and parcel of the metaphysical tradition which Heidegger is eager to destruct. Paradoxically enough, in opposing the reading of Heidegger's concept of formal indication in terms of scientific realism, Kochan brings the framework of "mind – the world out there" to bear on his analysis of Heidegger's position. He engages himself in the game of situating this position in realism debate. Yet Heidegger's difference between existence and essence (along with the difference between *praxis* and *theoria*) is not to be recast in the framework of this debate. The doctrine of minimal realism presupposes the primacy of a "minimal epistemic attitude" persisting in all modes of being-in-the-world, which contradicts the tenets of hermeneutic phenomenology. Granted that the epistemic subject-object relation is "always already" situated in a meaningfully articulated medium, it would be non-sequitur to insist on the existence of an independent world-structure that is beyond or behind the world-horizon which, according to existential analytic, transcends all entities that are ready-to-hand or present-at-hand. By admitting that physical reality is something like a substratum upon which meaning is imposed, one commits to a version of metaphysics of presence. The meaningful articulation of the world in the modes of being-in-the-world has nothing to do with transforming a pre-existential reality-as-presence into a meaningful reality within existence. The meaningful articulation "always already" takes place within the modes of existence, i.e. within the world already projected as a horizon of understanding.

There is no physical reality – however minimal it could be – preceding the physical-reality-within-the-world. The assumption that a pre-existential substratum is made meaningful by existence runs against the tenets of hermeneutic phenomenology. However, the minimal realist has to admit nolens volens the transformation of such a substratum into a meaningful reality (cum the sort of "non-Kantian Kantianism" that grounds it) in order to sustain the "minimally metaphysical" claim of his position.

Kochan is willing, as it were, to retain a residue of the subject-object polarity as independent of that ecstatic unity (of Dasein) with what is ready-to-hand within-the-world which hermeneutic phenomenology advocates. He does not provide us with clear arguments for this move. Having in mind, however, that Kochan tries to reconcile his Heideggerianism with Bloor's version of

SSK, it is my contention that the arguments are to be found in Bloor's (1999) celebrated counter-criticism of Latour's criticism of cognitive sociology of science. The author of the explanatory principle of symmetry argues that rejecting the individualistic and transcendental versions of the subject-object (society-nature) schema does not imply rejecting it in every sense. Thus, the Strong Program suggests an anti-individualistic and naturalistic framework of the schema for analyzing knowledge. More specifically, in opposing Latour's "generalized principle of symmetry" – nature and society are to be seen as co-produced – Bloor denies that his program seeks to explain nature in terms of society. The task consists rather in explaining the beliefs about nature by society, granted that society itself is part of nature. Thus, the door of combining anti-individualistic approach to epistemic beliefs with (a second-order) naturalism is open. Accordingly, Bloor's version of subject-object schema becomes allegedly justified. What becomes also justified in this plea for the original version of the principle of symmetry is the central position of the social actors as a point of reference in forging causal explanations about the social genesis of scientific beliefs. I will show a bit later that it is the impetus toward explanation that drives Kochan to accept Bloor's arguments for an irreplaceable (minimalist) subject-object schema.

Along with preserving a residue of subject-object dualism, Kochan tends to support (in a quite non-Heideggerian manner) another kind of dualism – that between individualism and social holism as this dualism is advocated in the traditional philosophy of the social sciences. The fact that Kochan is not willing to sacrifice not only the primacy of the subject-object dualism, but also traditional sociological dichotomies is typically epitomized in a critique of my view of trans-subjectivity he launched elsewhere. (See Kochan 2014) In reading my conception in terms of an attempt to defend the individual from the repressive manipulation of anonymous sociality (*das Man*) qua a political agent, he again imputes claims to my position that are completely alien to it. No doubt, *das Man* is not an agent. But it is not a social essence external to the individual existence as well. *Das Man* as horizons of anonymous sociality within which the individuals are "born" and "thrown" is inherent in the individual existence. The individuals are (in their inauthentic existence) thrown in the practices and possibilities of the anonymous sociality. I am not trying to identify the anonymous sociality as a "political agent" in a repressive society. It is my contention that the manipulative force of the anonymous sociality is even more effective in mass-democratic societies without oppressive regimes.[16] Dasein is the hermeneutic-circulative unity of mineness (*Jemeinigkeit*) and the horizons which by transcending mineness situates it in the world of being-with (*Mitsein*). (This unity as a kind of situated transcendence is captured by the ontological concept of trans-subjectivity which again does not amount to the sociological concept of inter-subjectivity.)

The emancipation from the horizons of anonymous sociality as a resolute "wanting-to-have-a-conscience" does not imply a liberation from trans-subjective horizons. The authentic existence of the resolute Self is also situated in such horizons. Kochan splits Dasein's trans-subjective unity, and blames me subsequently for reifying the anonymous sociality. Actually, due to this splitting he reifies both personal individuality and sociality. His "nuanced reading" of Heidegger brings into play a wrong sociologization of Dasein. Following the reception of the Durkheimian methodology in the cognitive sociology, Kochan reads Dasein dualistically. The Durkheimian roots of the Strong Program are palpable in the way in which this program preserves the figure of *homo duplex*, the co-presence of individual person and sociality, in all of its objects of inquiry. Kochan tacitly admits this co-presence when he asserts in the present volume that "scientists must be studied phenomenologically, in terms of their being-in-the-world … through which they come to make sense of … the subject matter of their investigations." But hermeneutic phenomenology of science is not a kind of phenomenological (or interpretive, or whatever) sociology. It does not study scientists. It deals with the trans-subjective being of scientific inquiry. "Scientists" are important only to the extent in which they participate in the constitution of this being. Kochan wrongly assumes that the modes of being-in-the-world are designed by social subjects.

Against the background of Kochan's minimal-realist reception of Heidegger, I will try to make sense of his allegations. He blames cognitive existentialism for (1) not considering science based upon mathematical projection of nature as a mode of existence, which entails a distortive construal of this projection as a "theoretical phenomenon"; (2) misinterpreting the existential conception of science as assuming a kind of mathematical essentialism; (3) ignoring the existential integrity of practices, and accordingly, not taking into account Heidegger's "phenomenological description of the change-over in scientific experience and its relation to the mathematical projection". I will meet these objections by raising a series of counter-objections. (In contrast to his [Kochan, 2012] earlier estimation of cognitive existentialism, Kochan presents now my reading of Heidegger as a continuation and a radicalization of Joseph Rouse's interpretation of the existential conception of science. He writes in particular that "Ginev largely accepts Rouse's interpretation" of Heidegger's account of scientific research. Though my whole work documents that this is not the case, I should like to supply my rejoinder with a brief remark in this regard. Over the years I have criticized Rouse's interpretation of Heidegger. (Ginev, 1997b, 2000, 2005) Since 1986 I have accepted Kockelmans's (1985) interpretation of the existential conception of science as grounding my search for a hermeneutic alternative to the objectivist philosophy of science. It is strongly based upon (a) the ontological difference and (b) the holistic-semantic approach to scientific knowledge. It is hard even to find a boundary point between Rouse's and Kockelmans's interpretations. Strangely enough, Kochan does not refer at all to Kockelmans's classical work in his efforts to bring Heidegger's philosophy to bear in science studies.)

A central motif in Kochan's line of argument is that existence is not to be explained by practices, but vice versa.[17] But is it possible (and admissible) to approach existence in terms of explanans entitled to make intelligible the phenomena which supposedly take place in it? In *Being and Time* there are no explanatory scenarios but constitutional analysis and hermeneutic interpretation. Does Kochan reading of the existential conception go so deep as to integrate explanatory scenarios in the existential analytic? If Kochan wants to "improve", or "revise", or "enrich", or "extend" hermeneutic phenomenology by integrating in it (perhaps sociological) explanations, then the discrepancies between our methodological positions are really insurmountable. (He does not aim at complementing the constitutional analysis of the scientific mode of being-in-the-world with empirical explanations. Strangely enough, Kochan aspires to expanding the explanatory power of what he counts as phenomenology of scientific inquiry.)

It goes without saying that cognitive existentialism (like any other version of hermeneutic philosophy of science) approaches science as a mode of existence based upon the pre-objectifying being-in-the-world. Does cognitive existentialism explain existence through practices? An affirmative answer would have to presume that the interrelatedness of practices is considered in this doctrine as a present actuality that can be schematized appropriately to provide (say, teleological) explanatory principles. (When Kochan says that cognitive existentialism explains existence by practice, he has in mind, I suppose, that the doctrine he criticizes is something like a functional-teleological theory of action.) Since the interrelatedness of practices has a being only in possibilities, the schematization in question cannot take place. Some additional methodological clarifications are in place here.

Regardless of the type of (causal, functional, etc.) explanation aimed at, each explanatory strategy requires logically and methodologically the demonstration of a foundational presence. Conceptualizing this presence builds explanans. On cognitive existentialism, practice (and the interrelatedness of configured practices) are not – as Kochan tries to suggest – "a new explanatory concept" admittedly alternative to that of conceptual framework. The interpretive-phenomenological way of entering into hermeneutic circles of the constitution of objects of inquiry and knowledge in scientific research is what cognitive existentialism aims to attain. Roughly, scientific research is a texture of practices that are within horizons of possibilities projected by the very interrelatedness of these practices. On the doctrine of hermeneutic realism I defend, because scientific inquiry is a mode of being in practices that project possibilities whose appropriation shapes the interrelatedness of these practices, the reality thematized and investigated in scientific inquiry is within the circularity of practices and possibilities. (Like scientific inquiry itself, the reality in scientific inquiry has exclusively a potentiality-for-being.) Accordingly, this reality is not to be disentangled from the hermeneutic situations in which it becomes thematized and investigated.

Thus considered, scientific inquiry does not have a being as presence that can be conceptualized as explanans. Its potentiality-for-being is characterized by several kinds of hermeneutic circularity that are fore-structuring the articulation of scientific worlds (including the worlds of technoscience). The task of cognitive existentialism is to enter in the right way the hermeneutic circles of reading-that-textualizes – the textualizing reading by means of scientific practices as readable technologies. In adumbrating the non-explanatory profile of cognitive existentialism, however, I am realizing that there is a much more general question to be asked: Why should one (who confesses his commitment to phenomenological-constitutional analysis) become engaged with explanations at all? Initially, I thought that the only reason for Kochan's formulation that "forms of existence explain practices" is an undesired and (perhaps) accidental lapse. A careful reading of Kochan's papers, however, convinced me that he is really seeking for (causal) explanations in which existence plays the role of explanans. For instance, he firmly believes that "the new regimes of epistemic practice must finally be explained in terms of changing modes of existence."

Where Kochan's "explanatory impetus" does come from? An old debate over the reception of Wittgenstein's philosophizing of meaning-as-use gives perhaps the answer. The debate goes back to Bloor's (1992) accusation of ethnomethodologists. He blames them for an incorrect analysis of what "rule following" is. More specifically, he severely opposes the view that the empirical investigations of scientific practices based on Wittgenstein's work are to be non-explanatory initiatives. Rule following in practice is intrinsically determined by social factors. Following this claim, Bloor makes the case that explanatory notions (borrowed from social theory) like consensus, convention, or socialization refer to events and processes that are not extrinsic to the rule-following practices. Since they are constitutive – which in in this case is tantamount to being determining factors – of these practices, their conceptualization enables one to devise sociological explanations that are not externally imposed upon what becomes explained. Holding that the internal relation between rule and its contextual application is a social relationship amounts to postulating that the rule-following practices are to be explained by social essences. Consequently, these practices are to be only made intelligible by means of explanatory sociology. (On Bloor's account, the kind of explanation suggested by the "naturalism about the social" is to be opposed to the anti-explanatory enthnomethodological description.)

It is not my job here to appraise which account of Wittgenstein's concept of rule following – the construal suggested by the traditional SSK or the account advocated by prominent ethnomethodologists – is the correct one. My point is only that if one accept the explanatory position – as Kochan does – one ought to admit that consensus, socialization, ethos, tradition, convention, etc. are not constituted in the hermeneutic circularity of interrelated-practices-in-horizons-of-possibilities and contextual (and rule-following) actualizations of possibilities. Attributing to these events and process explanatory functions means that they are isolated from the hermeneutic circularity in question. This runs against the intention of studying the rule-following epistemic practices of scientific research as constituted by the mode of existence characterizing modern science. The basic assumption of the explanatory sociology of scientific knowledge goes as follows: There are social essences that – being constitutive of science, and accordingly, playing methodologically the role of explanans – determine causally all features of science-as-a-mode-of-existence. By subscribing to the sociological-determinist strategy, Kochan shows once more his indebtedness to Bloor's explanatory naturalism. Yet in contrast to Bloor, who does not profess a commitment to existential phenomenology, Kochan is doomed to face serious troubles in matching explanations (in which existence is explanans) of scientists' practices with phenomenological analysis associated with the claim that scientific inquiry proceeds in hermeneutic circularity. (Not by chance he takes Edgar Zilsel's historical sociology of modern science to provide explanation of science's existential genesis. No doubt, hermeneutic phenomenology of scientific inquiry and historical sociology of science have to work in concert in many cases. But this does not imply that the former is to be modified in an explanatory enterprise. Zilsel provides us with superb sociological explanations of how the cooperation of various social strata has been contributed to the institutional rise of modern science by enabling "the fusion of mathematics and scientific experiment". Yet his historical sociology has nothing to do with the problematic of science's existential genesis – a problematic that is not to be formulated at all within the framework of Zilsel's investigations. His comparative-sociological-explanatory method is highly inventive and helpful in identifying the historical origins of modern science. *Pace* Kochan, however, this method is completely irrelevant to identifying the "existential origins" of science. It deserves mentioning that Zilsel was a strong antagonist to phenomenology. He was in particular preoccupied with demonstrating that in no methodological context can the phenomenological description serve as an alternative to causal explanation.[18])

B

Kochan's nuanced reading of the existential conception of science is a source of further embarrassment to me. He imputes to Heidegger's conception the problem of a choice between theory and practice, asserting that this is a choice between "two distinct epistemic modes of the same underlying phenomenon of existence". What Heidegger (1962, p. 238) holds, however, is that praxis and theory are "possibilities of being" for that ontically distinguished being (*Seiende*) whose being (*Sein*) is to be ontologically defined as "care".

Indeed, one can find in Heidegger's later work the formulation that *theōria* is "the consummate form of human existence." (Heidegger, 1977, p. 164) But the passage, in which the formulation appears, discusses what Greek experience understood by *bios theōrētikos*. Notoriously, Heidegger (1977, p. 166) decisively contrasts Greek *theōria* with modern concept of theory: "The theory that modern science shows itself to be is something essentially different from the Greek *theōria.*" In the same paper ("Science and Reflection") Heidegger is very clear about "the essence of modern science". His paramount claim is that the characteristics of science-as-research determine the reality in which man of modernity moves and attempts to maintain himself. The essence of science is to make what it objectifies to come to presence. This essence becomes manifested in the way in which modern science is "the theory of the real". The real in the modern age is what brings its presencing to stand in objectness. This is the way – to reverse the formulation – in which the real presences (or shows itself) in modernity. It is science that determines this destiny of the real in modernity. The whole modern age is, as it were, encompassed by a global scientific theory that makes any emerging phenomenon within the scope of modern experience to fit "into the normative objective coherence of the theory".

Here is a point that I would like to make in advance. In countering to a certain extent the Heideggerian picture of science-modernity relationship, one is to state that not modern science's quest for objectivity but the modern epistemic attitude (which historically precedes scientific objectivism) is what brings into being the holding-sway of presencing in terms of objects. (When Francis Bacon, for instance, replaced the humanist ideal of individual glory with the ideal of controlling nature by means of scientific knowledge – as this is beautifully explained by Zilsel [2003, pp. 17–20] – he made explicit the ethos of modern epistemic-instrumental attitude and not an ideal that allegedly resides in early modern science.) Modern science only incarnates a form of this attitude in the way in which it objectifies the world. Following this claim, one would become engaged in an essentially different (from Heidegger's one) cultural-historical scenario about the role of science in modern age's way of revealing/concealing. There is a shortcut in Heidegger's line of argument concerning the relationship of scientific objectivity and modernity's technological instrumentalism. It is my contention that this shortcut is informed by Heidegger's earlier "mathematical essentialism" which I am going to address later in this rejoinder. But let me return to my criticism of Kochan.

The problem with Kochan's – allegedly Heideggerian – contradistinction is that if practice and theory are two distinct epistemic modes, their description might be conducted entirely in an ontic way (e.g. in terms of intellectual history or cognitive sociology). Heidegger's approach is designed to prevent one from such reductionism. Defined as "possibilities of being" praxis and theory are characteristics of each mode of existence. In their potentiality-for-being, praxis and theory are first and foremost ontological phenomena (i.e. phenomena projected on horizons of existential possibilities). Only the contextual actualization of certain possibilities would make them ontically identifiable epistemic forms. Phrased differently, only in their actualized presence they appear as such forms. Kochan's reading, which ignores the ontological difference, is obviously excessively nuanced since it enriches the existential analytic of scientific objectification with new formulations. The assertion that theory and practice are "epistemic modes of existence" is not only not to be found in Heidegger's work, but it even cannot be recast in terms of hermeneutic phenomenology. The less are practice and theory "different determinations within the same experiential field of scientific existence."

In analyzing the final step in the existential genesis of scientific objectification, Kochan concludes that this is "the movement from non-deliberative, non-propositional know-how to deliberative, propositionally structured know-that". *His conclusion is quite the opposite of Heidegger's position.* For Heidegger, deliberation and circumspective deliberation (or, as I prefer, deliberative circumspection) are precisely those existential characteristics of practical experience which become annihilated by scientific objectification. To put it in Heideggerian terminology, deliberation (*Überlegung*) is the specific way of interpreting what is ready-to-hand circumspectively. Deliberation (that operates without epistemic representation) is one's self-referring survey (and in this sense, the primary and most elementary form of reflexivity) of one's being in the practical work-world. On the respective passage from *Being and Time*, deliberative circumspection "receives its 'light' from the potentiality-for-being", thereby making the contextual concern of practical experience to exist as care. (Heidegger, 1962, p. 410) Deliberative circumspection "makes present" by not leaving room for a non-contextual (i.e. beyond any circumspective concern) presence. (In my rejoinder to Gironi I stated that this capability of preventing one from admitting a non-contextual presence correlates with intrinsic reflexivity of deliberative circumspection.) For Heidegger, scientific research operates without "circumspective making-present" because it is freed from deliberation. Accordingly, the epistemic progress of science does not make it "more deliberative" as Kochan assumes. It makes science more procedurally effective in objectifying thematization, i.e. in "releasing from contextual confinement" thematic regions as "pure manifolds" of formally determinable positions. (John Macquarrie and Edward Robinson translate *Mannigfaltigkeit* as "multiplicity". Another translation, closer to the technical meaning of the term in mathematics, is "manifold". Not by accident Heidegger uses here this mathematical term. He borrows it from Husserl in whose philosophy it plays a prominent role. Notice that Husserl's mathematical background is closely related to the theory of manifolds. In his earlier work Husserl makes uses of the term "manifold" in non-mathematical contexts, retaining, however, its general mathematical meaning as a system of relations between elements determined by axioms. Stephania Centrone [2009, p. 185–196] pays considerable attention to the concept of "constructible manifolds" in her study of Husserl's philosophy of mathematics, nicely showing how its reinterpretation transmits ideas discussed in the context of *Philosophy of Arithmetic* to the context of *Formal and Transcendental Logic*. It is the – initially psychologistic, later anti-psychologistic – non-mathematical extension of the theory of manifolds that made possible Husserl's transition from studies into foundations of mathematics [as these studies are, in particular, informed by Weierstrass's later work] to phenomenology "as rigorous science". Heidegger admits in a pretty Husserlian manner that to each mathematical projection of a region there corresponds a spatial-temporal manifold. The latter here is to be understood as the class of all models of the [axiomatic] system of mathematically codified theoretical laws which govern the behavior of the region's empirical entities. There is an obligatory moment of invariance involved in the mathematical concept of manifold – the manifold is invariant under different descriptions. It is this invariance that in the structuralist-semantic conceptions of scientific theory becomes crucially tied with scientific objectivity. Heidegger follows to a great extent this line of reasoning, thereby paying tribute to a special view [and ideal] of scientific objectivity.)

On Heidegger's existential conception, science's procedural effectiveness involves (1) the effectivity of the releasing (of the pre-scientific entities that will be transformed into objects of inquiry) from their inner-worldly environmental confinement (*die Entschränkung der Umwelt*) through idealization; (2) the effectivity of schematizing the region of research as a manifold of spatio-temporal positions that remains invariant with respect to a group of transformations whereby the spatiality and temporality of deliberative circumspection become objectified as metric space and measurable time; (3) the efficacy of the complex quantification which involves the efficacy of measurements, computations and calculations of what (within the region of research) is mathematically schematized; (4) the effectiveness of deriving empirically testable consequences from the postulates (axioms) whose theoretical terms incorporate the variables in the (idealizational) mathematical schematization which on its part is governed by principles of invariance. (This is the procedure of constructing a "theoretically laden" empirical basis ["theoretical factuality"] which Heidegger – though not using post-empiricist terminology – discusses on several occasions.) Generally speaking, (1) – (4) depict a profile that is in line with the methodological tenets of deductive-nomological theorizing.

At this point, a short note concerning [3] is in place. Actually, this kind of efficacy expresses most typically what Heidegger calls "the mathematical projection of nature". Indeed, Heidegger assumes that the mathematical projection opens the horizon of possible research within the projected region. Yet there is in this usage no one of the connotations Heidegger has in mind when addressing understanding as an open horizon of possibilities. The metaphor of horizon when attributed to what mathematical projection opens is used in a quite peculiar manner: The mathematical projection opens the horizon of research by determining a priori what might be quantitatively studied. The mathematically fixed region determines the scope of possibilities that are horizonally open. Tentatively speaking, combining horizonality with a priori determinism is a kind of essentialism. The word "projection" might provoke a further confusion by compelling one to admit that there is a kind of continuation between the way in which *Dasein* projects its being upon possibilities and the mathematical projection. (Ginev, 2012)

C

In this section I am going to discuss an unnuanced facet of Kochan's nuanced reading of Heidegger. This reading neglects the issue of science's epistemological identity as it is adumbrated in the existential conception. True, Heidegger distinguishes between existential and logical conception. (In relating the logical conception to the issues of empirical validity of logically coherent theoretical constructions, he has a historical priority over Hans Reichenbach in introducing the concept of the "context of justification", *Begründungszusammenhang*.[19] The expression *Begründungszusammenhang* is in this case synonymous with the expression *Rechtfertingungszusammenhang* in the sense in which the latter will become in the 19230s established in logical empiricism. Reichenbach introduced the context-distinction [when dealing with – in his words – "problems of a deductive character"] in *Experience and Prediction* [1938]. He did this with the intention to prevent one from confusing mathematical with psychological relations. By the same token, Heidegger delineated the context of justification ten years earlier with the intention to demarcate the problematic of phenomenological analysis of science from the issues of science's logical reconstruction which is addressed by the "logical conception". He contrasts "deductive justification" [*Begründung*] with constitution as "existential genesis". Unfortunately, John Macquarrie and Edward Robinson omit the term "context of justification" in their translation of *Being and Time*. Instead, they use the expression "interconnection of true propositions", which does not do justice to the justificatory-normative-evaluative function Heidegger assigns to the logical conception of science.)

Now, one should keep in mind that the way in which Heidegger contrasts the phenomenological analysis of science's theoretical attitude with the logical conception of science is only a preliminary distinction in the course of unfolding the existential conception of science. The "logical conception" is not the one which reconstructs correctly the cognitive organization of scientific research as it is envisioned by the existential conception. In this regard, another theory of science's cognitive organization is in need to complement the existential conception. Kochan does not take into account this motif, and his way of coping with Heidegger's view about the existential genesis of science's logical structure is disappointing. Skipping the analysis of the holist (non-propositional) approach to theoretical objectification in the existential conception damages the whole discussion of "Heidegger's philosophy of science". The "logical conception" is not only incapable to go beyond the "context of justification", but it is ill-equipped for addressing the structure of theorizing within this context. In fact, the logical conception approximates the image of scientific theory qua an axiomatic system that obtains partially an empirical interpretation through rules of correspondence. (It is the implicit criticism of the propositional view of scientific theory adumbrated in the existential conception that would prepare Heidegger's later sketchy criticism of logical empiricism.)

Against the background of these considerations, the question one has to ask is what kind of epistemological image of science the existential conception of science does imply. My contention is that though missing the appropriate terminology, Heidegger tends toward a semantic conception of theoretical objectification. The semantic conception suggests a holist view about scientific theory. From the perspective of this view, the "logical conception" is criticized for its inability to deal with the unitary structure of scientific theory. The logical conception is only preoccupied with the syntactic relations among particular propositions formulated in the framework of a theory. (The logical empiricists who in the early 1930s adhered to this propositional approach will later realize that it is closely connected with the troubles provoked by the debates over the protocol sentences.) In addition to the paramount appraisal of the logical conception's irrelevance to the way of handling science as a mode of being-in-the-world, there is in the existential conception a tacit proposal for a holistic approach to the structure of scientific theorizing that through semantic means can avoid any appeal to elementary factuality as pure presence beyond theorizing. *The holistic image of theoretical objectification which the existential conception implies has to be kept in mind when one tries to estimate Heidegger's concept of the mathematical projection of nature in light of essentialism-existentialism contradistinction.*

The arguments for linking the existential conception to a kind of semantic-structuralist approach to theorizing are to be found in Heidegger's description of what he calls a "delineation of a region" of scientific research. He assumes that the region is defined by the totality of empirical models satisfying the theory's laws *en bloc* (as a unitary predicate amenable to interpretation in various sets of models). Again, Heidegger never made this assumption explicit because he lacked the germane philosophy-of-science's language for this purpose – the language of the non-propositional view of theory's structure. Notoriously, by treating scientific theories as set-theoretical (extralinguistic) entities, this view avoids some major difficulties encountered by the "logical conception" and related to the infamous distinction between theoretical and observational language. The sets of models that interpret the mathematically codified theoretical laws brings the whole of what is – according to the existential conception – released from its environmental confinement as the semantic whole of a region of research. Conversely, the region is to be envisioned by means of its theory's semantic interpretation. It is this interpretation that reveals the sense in which the region has become mathematically projected. On this account, the mathematical projection is at once the syntactic formalization of a region's theory (the mathematical codification of its laws) and the range of the theory's possible semantic models. Thus considered, the view of science's cognitive structure implied by the existential conception puts the semantic interpretation first. To be sure, there is also a room in this conception for a kind of hermeneutic interpretation of what becomes conceptualized (and formally codified) – the interpretation of the theoretical objectification within the horizon opened by the mathematical projection. Paradoxically enough, however, this horizon opens a limited leeway of possible theoretical and empirical articulation of the objectified region, without becoming an open horizon. To reiterate a previous claim, it is a horizon determined (and constrained) by the mathematical projection. Accordingly, the semantic interpretation of the region's theory determines the limited possibilities for doing research that articulates further the region being already mathematically projected.

But what Kochan's reading of the existential conception of science misses most dramatically is the construal of the conception's intent (and the role it plays in the whole plot of *Being and Time*). Heidegger discusses the mathematical projection of nature with regard to the different regimes of temporalizing of temporality. The primary goal of the existential conception of science is to reveal the temporality of scientific objectification. The existential conception (like all doctrines of the existential analytic) is a kind of transcendental quest: The "existential condition of the possibility" for being-in-the-world by doing scientific research must be sought – so Heidegger's (1962, p. 404) argument goes – in the modifiability of the temporalizing of temporality. At issue in the existential conception is a certain modification of this temporalizing that enables *das thematische Erfassen*, thereby providing the existential condition in question. Scientific objectification based on the mathematical projection is of prime importance of the whole existential analytic since its extreme regime of temporalizing of temporality – the absolutizing of the present beyond any circumspective making present – elicits the existential transformation of circumspection into a "mere representation". This is a transformation that inaugurates – in Heidegger's later view – the (replacement of the world by) "World-Picture". With respect to the way in which the temporalizing of temporality is taken into account (or ignored), I introduced elsewhere a distinction between sophisticated and trivial readings of the existential conception. (Ginev 2011, p. 2–11) Kochan's description of the four steps does not cross the threshold of the trivial reading.

There is a more general point to be made: Provided that Dr. Kochan completely ignores the dimension of temporality, I am not able to comprehend at all his constant appeal to phenomenological analysis. The linking of the study of meaning constitution with the investigation of the modes of temporalizing makes this analysis resistant to any attempt at its objectivist-naturalist reduction to empirical (e.g. socio-psychological) studies that take the production of meaning to be accomplished within the objective presence of the ontically identifiable time. Anticipating my further considerations, I would state that the decisive argument for categorizing Heidegger's view of mathematical projection as a kind of mathematical essentialism concerns exactly the temporality of scientific objectification.

Had Kochan paid attention to the context in which the existential conception had been developed, he would have seen that Heidegger ascribes to scientific objectifying thematization – the scientific way of "making present" – the same kind of temporalizing which he attributes to that kind of hypostatizing the present which leads to metaphysics of presence. This anticipates Heidegger's later (after *die Kehre*) critique of science. (Kochan refers in his paper to some developments of the concept of mathematical projection in Heidegger's later philosophy. He believes that these developments should provide additional support of his reading of the concept in terms of "mathematical existentialism". It seems to me, however, that he is not concerned with the important caesura which the concept has been undergone in Heidegger's *Denkweg*. There is no mere continuation in the ways of working out the concept. Some tendencies alluded in the existential conception become strengthened and widely extended in Heidegger's late philosophy. As a result, the concept of mathematical projection is entitled to play a significant role in the characterization of modern epoch as an epoch that in a specific manner reveals and conceals being in the "epochal history of being". Yet, however substantial is the concept's transformation, it still preserves the kind of mathematical essentialism about scientific inquiry which the existential conception has been launched. Notoriously, in Heidegger's later philosophy the critique of scientism – the cooperative unity of instrumental-technological and scientific rationality – rests on the claim that the fundamental event of modernity is the conquest of the world as picture. The picture is the objective representedness of all entities that can be exploited and technologically manipulated. Heidegger (1977, p. 135) blames precisely the capability of modern science to make the world as picture ready to be conquered – science that "simply chases after results and calculations". He stresses that science-as-research is the main pathway upon which the modern age "rages toward fulfillment of its essence." This critique presupposes an intrinsic instrumentalization of science that transforms scientific research into *Betrieb*. Heidegger admits, however, that the instrumentalization – science is only for the sake of serving instrumental functions of modernity's "gigantism" – is a natural tendency determined entirely by the "essence of modern science". The latter participates in "accruing of power by power for its own overpowering". In rejecting that modern science has an essence and in line with arguments adduced earlier, I claim that Heidegger confuses the nature of scientific research with the political [mis]uses of science. [Ginev, 2013a])

The view of mathematical projection becomes radicalized after the turn in Heidegger's thought as a view of "the metaphysical projection of the thingness of the things". The mathematical projection becomes now a *dimension of the (modern) history of being*. This projection is the essence of modern science and not a particular practice (or a bunch of practices) of scientific research. The mathematical projection "sketches out in advance the manner in which the knowing procedure must bind itself and adhere to the sphere opened up" whereby the totality of the entities of knowledge becomes the world conceived and grasped as picture. (Heidegger, 1977, p. 118) (It seems to me that Kochan wants to create a painting of Heidegger's highly friendly attitude toward science. A feeling of sincere optimism emanates from this painting. Science has its own "pattern of existence" that promotes the "desire to explore" and motivates "the development of new fields of material practice". Reading Kochan's Heidegger leaves the impression that the historical figure of Heidegger is to be received as a forerunner (and even a pioneer) of science and technology studies. This interpretation – however it might be valuable with regard to Heidegger's reception in SSK and STS – misses the basic critical intention of the thinker's dealing with science as part and parcel of his critique of modernity. I am not saying that there is no interesting critical impetus involved in the science studies being mentioned. Yet this impetus is of essentially different sort as compared with Heidegger's critical agenda.)

D

I already mentioned passim some of the arguments for the thesis that Heidegger's existential conception of science (and most of all the concept of mathematical projection of nature) rests upon what I call "mathematical essentialism". Let me now spell out in more details this expression. Under the heading of "essentialism" are to be included the doctrines which admit that the potentiality-for-being of a mode of existence – in the most extreme case, the totality of existence – is causally (or functionally, or normatively, or structurally, or teleologically, or transcendentally) determined and constrained by an essence that has the character of a pure presence (like the world-out-there, the biological substrate of human life, the "memes", a normative code, a formal structure, fixed goals, trans-contextual values, static a priori conditions, or "the transcendental signified"). Put alternatively, a kind of essentialism is advocated when it is assumed that the existence's potentiality-for-being is determined (restricted, constrained, normatively regulated) by something to which the status of a pure presence – actuality without an *open and potentially infinite* horizon of possibilities – is assigned. By imposing compellingly a determining force upon a mode of existence, essentialism rules out the ontological primacy of horizonality and interpretive openness regarding this mode. This is why the latter becomes deprived of a genuine potentiality-for-being. The mode of existence is pressed on possibilities that are a priori determined and constrained whereby existence becomes an epiphenomenon of the determining essence. The existence is amputated from its capability to project its own open horizon of possibilities (as the horizon of the particular mode of existence's self-understanding), and to constitute constantly its own identity through the appropriation of these possibilities.

Conceding that the postulated presence/actuality is a historically invariant essence does not belong to the obligatory conditions for defending essentialism. The potentiality-for-being of a given mode of existence might be determined by an essence not "for ever and ever". The relation of determination can take place only in a particular historical situation, or the role of determining essence might be played by contextually valid (and in this sense contingent) norms and values, or by "factors" operating in a certain cultural setting during a definite period, or by a historically limited normativity of a culturally specified mentality. In all of these cases, the determining force imposed from without on a mode of existence is historically and culturally relativized. Relativizing essentialism leads to a diversification of its forms, but not to its undoing. Thus, the advocacy of essentialism (as well as the struggle for overcoming this position) are undertakings independent of the contradistinction of relativism and universalism. To sum up, one can be a good relativist, defending at the same time a position of strong essentialism.

Heidegger's "mathematical essentialism" consists in the view that a particular mode of existence – the being-in-the-world-by-objectifying-the-world-thematically which is procedurally emancipated from the "average everydayness" – is determined by a projected mathematical essence. On the existential conception, the mathematical projection replaces the open horizon of deliberative circumspection. Although it looks like a finite horizon for those who are appropriating the possibilities in a particular context, the horizon of deliberative circumspection transcends (as the world-horizon) each particular context of dealing with what is ready-to-hand. (This horizon is not an actual infinity of possibilities, but rather a trans-finity in the sense of being predicated on the capability to transcend constantly its own contextual finitude. Cognitive existentialism advocates the view that the horizons of scientific inquiry as a mode of existence retains the character of potential trans-finity. Heidegger's existential conception denies this view.) On this conception, the mathematical projection succeeds in de-worlding "the aggregate of present-at-hand", which is a prerequisite for having what Heidegger (1962, p. 413) calls "the theme" of scientific research.

According to Kochan, Heidegger cannot view the entities subjected to objectifying thematization as having been de-worlded, because he still attributes to them "world-points". This claim does not make sense in the existential conception's framework. (More precisely, it is a claim that runs absolutely against the intentions of the author of *Being and Time*.) Heidegger is quite clear about what he means by a spatio-temporal position (or a "world-point"). It results from the replacement of the horizon of temporalizing (and spatializing) in the appropriation of possibilities which constitutes meaning by a pure mathematical manifold. Due to this replacement the place (as a characteristic of making-room within the existential spatiality) becomes a matter of indifference. On Heidegger's account, it is completely wrong to regard (as Kochan is willing to do) the world-point as a minimal manifestation of the world as horizon. Transforming the world into a manifold of world-points is the quintessence of de-worlding by means of the mathematical projection.

To put it lapidary, the mathematical projection determines a priori the way of deploying, the design, and the course of execution of scientific practices, but this projection itself does not consist of practices. (Kochan seems to be inspired by Heidegger's claim that "the theory" is not without its own praxis. This sounds as if the mathematical projection is not to be exempted from the practical milieu of scientific research. Heidegger cites in this regard a wide spectrum of scientific practices between the poles of laboratory experimentation and mathematical operations with theoretical entities. However, Kochan's enthusiasm is to be cooled. Heidegger takes the scientific practices he mentions to be only a continuation of the "theoretical seeing".) In Heidegger's work mathematical essentialism is epitomized by four (more or less unfolded) lines of argument – an ontic, a transcendental-epistemological, a methodological, and a metaphysical one. Through each of them the mathematical projection is identified as determining essence.

(1) On the ontic line, the mathematical projection determines the possible entities that can be constituted as objects of inquiry in a delimited region. These entities are projected in their totality as a "constant presence-at-hand" (*ein ständig Vorhandenes*) that is "uncovered beforehand". (The projection is responsible for the transformation of the pre-thematic readiness-to-hand into a thematic presence-at-hand.) The ontic determination (of the possible objects) achieved by the mathematical projection refers in the first place to the (fixed number of) idealized theoretical objects characterized by quantitative variables. (All of the empirically identifiable objects are particular configurations of the measured values of these variables.)

(2) On the transcendental-epistemological line, the mathematical projection determines the possible epistemic regimes of scientific research. The mathematical projection opens a (methodological) horizon of doing research within a delimited region, but it itself is not a horizon (in the phenomenological sense). The mathematical projection is accomplished by means of a formalism that is applied not in an ad hoc manner to particular phenomena in a particular situation but reveals the essence of the delimited region as an invariance specified in terms of spacetime symmetries. The epistemic regimes of research are designed in accordance with these symmetries. Through such a formalism only can the projection disclose a region a priori, thereby making it ready for empirical research. The "state of being" of the region is thus determined. (One way of characterizing the horizon which is opened by the projection is with regard to the leeway of possible formal operations with the theoretical objects as this leeway is determined by the respective formalism.) Projecting a formalism determines scenarios of how to create factuality (theoretically laden facts) within the epistemic regimes of scientific research. (Once the region becomes mathematically projected, there are no longer "bare facts" [*bloße Tatsachen*]. All facts are singled out within scenarios determined by a formalism.)

(3) The methodological line is related to the primacy which Heidegger ascribes to the "theoretical seeing" in scientific research. The existential genesis of science ought to be approached "in accordance with the priority of 'seeing'." (Heidegger, 1962, p. 410) The circumscription of "the theme" of scientific research de-temporalizes deliberative circumspection of the pre-theoretical experience, thus transforming this circumspection into "theoretical seeing". Here again at issue is the determined and constrained horizon opened by the mathematical projection. This time it is to be regarded as a methodological horizon for discoveries (of facts, measurable entities, relations between independent and dependent variables, empirical laws, theoretical laws,

formalizable structures, etc.) by means of scientific practices. (The methodological horizon opened by the mathematical projection is to be defined as a limited leeway of possible methodological maneuvers within the objectified region.) The mathematical projection determines the methodological horizon of possible discoveries within a delimited region of research. To reiterate, in *Being and Time*, "seeing" enabled by the mathematical projection of nature results from a transformation of circumspection that annuls the very circumspection – a view that supplements the claim about the constant presence-at-hand projected in the research process. Thus established, the theoretical seeing "regulates" all procedures and practices of scientific research. (This is why Heidegger considers "all practical manipulation in the sciences" as a continuation of theory by other means.) Cognitive existentialism does not rebut that there is a transformation of (pre-scientific) circumspection which elicits a de-contextualizing "theoretical seeing". Yet the latter resides in (and is contextualized by) a new kind of circumspection embedded in scientific practices that annuls each alleged presence-at-hand in the research process.

(4) The metaphysical line of argument for the essentialist interpretation of the mathematical projection is chiefly developed in Heidegger's later philosophy. It is a line unfolded in terms of the epochal history of being. After the *Kehre* Heidegger's interpretation of the concept revolves around the claim that because mathematical physics as a pure theory "sets nature up to exhibit itself as a coherence of forces calculable in advance, it therefore orders its experiments precisely for the purpose of asking whether and how nature reports itself when set up in this way." (Heidegger, 1977, p. 21) He holds that the challenge to reveal nature as a resource for possible exploitation already holds sway in science's cognitive structure. Accordingly, science's epistemological objectivity is rooted in the essence of modern technology. It is the nexus of science's objectivity and technology's exploitive instrumentality that made Heidegger's later philosophy especially popular among critical theorists of various sorts. The destiny of modern man is determined by this nexus. An "unreasonable demand" is put to nature not by the large-scale modern technology, but by the implicit interest in technological colonization which is entangled in the structure of objectifying science. Since the early 1960s this claim has become a magic formula for criticizing scientism from different perspectives. (Ginev, 2013a) The modern mathematical-experimental theorizing prepares first and foremost the way for the "essence of modern technology" (i.e. the way in which being becomes revealed by modern technology). In this sense, modern physics is proclaimed to be "the herald of Enframing". (For Heidegger, research is the essence of science just because research can be approached in essentialist terms. In the essay "The Age of the World Picture" he assumes that scientific research deals with a fixed *Grundriss* of natural events. Doing research is bound in advance to adhere to the projected plan. [Here projected plan is intentionally opposed to open horizon.] The "binding adherence" is for Heidegger the rigor of research. The complexity of the objects of inquiry disclosed by the projection determines the criteria of the rigor. In

enabling the procedural character of scientific research, the projection determines the essence of science. Science is a *sui generis* mode of existence – procedural dealing with fixed natural events – inasmuch as it is determined by what makes research the essence of science. In each particular region of scientific research this essence is projected in a characteristic manner. Due to this projection the region becomes sketched out in advance.)

The mathematical projection enables one not only to objectify thematically the world. The main function of mathematical projection in the modern epochal history of Being consists in transforming the world into ensembles of calculable and measurable entities. In this regard, Heidegger argues that modern natural science, modern mathematics, and modern metaphysics sprang from the same root of the mathematical.[20] (Indeed, after the *Kehre* the most important role in Heidegger's approach to science plays not the concept of the mathematical projection but the conception of "the mathematical". *Ta mathēmata* is conceived of as what is learnable, i.e. what can be grasped and appropriated. At the same time, the mathematical is the a priori knowledge about the thingness of the things – the things' essence which is already projected in advance with regard to any dealing with the empirical. The mathematical in modern epoch is the horizon of idealizing thinking that objectifies. [The objectifying idealization itself determines a finite region. But modern idealizing thinking is in the potentially infinite horizon of the mathematical.] Roughly speaking, the mathematical projection is that specification of the mathematical which takes into account the axiomatic structure of modern science's theorizing. The mathematical projection participates in the way in which being becomes revealed/concealed in the modern epoch of the history of being. Science's mathematical essence – "the mathematical" – gives one the clue for understanding modern metaphysics of presence. It is this essence that determines the manner of working with things. The claim that mathematics [as formalisms of scientific theorizing] is only a particular kind of the mathematical [*mathēsis*] cannot be formulated in the language of the existential conception. One can do justice to this claim only within the "epochal history of Being": In modernity the mathematical – through its neo-Platonic interpretation – becomes established in a certain direction that determines modernity's cultural-historical destiny.)

The mathematical projection is in a sense mystified both in the existential conception and in Heidegger's later philosophy. It grounds the whole existential genesis of science – making possible the theoretical attitude, objectifying thematization, and delimitation of regions of inquiry – without having such a genesis rooted in transformations of the pre-theoretical "average everydayness". But it has no empirical genesis too. Heidegger is not able to assert that the mathematical projection originates in empirical events that can be approached in a historiographical manner or by scrutinizing empirically the cognitive specificity of particular research processes. Such an assertion would contradict the phenomenological scenario of science's existential genesis. Thus, he implies that the mathematical projection is an event that can only be discussed in terms of the existential analytic by skipping the issue of its genesis. In *Being and Time* the concept of this projection is introduced – or better, postulated – as a part of a certain regime of temporalizing the temporality. Therefore, the mathematical projection is granted with a character that precludes at once its genealogical investigation and empirical study. I will argue later that attributing to it such a character is an unwarranted move.

Despite the irrelevance of the mathematical projection to events in the history of science or to cognitive practices of scientific inquiry, Heidegger's interpretation of the processes arising out of this projection promises to recast several subsidiary elements involved in science's existential genesis in terms of disciplinary discourses like history of science and cognitive sociology. The translatability of certain aspects of the genealogical (phenomenological) scenario into possible empirical programs is perhaps the greatest merit of the existential conception, and the reason of its vitality in today's science studies. Moreover, the translations of these aspects do not require – to take up a motif already discussed – "amendments", "revisions" or "extensions" of the existential conception (qua a phenomenological conception) by integrating in it explanations. (The translatability is not to be confused with such "explanatory revisions" of Heidegger's conception itself.) Paradoxically enough, however, the grounding concept of the genealogical scenario – the mathematical projection of nature – remains the only "untranslatable element". It does not correspond to empirical events, processes, acts or practices that can be addressed in cognitive-sociological, social-historical, cultural-historical, socio-psychological, etc. terms because it is not scrutinized existential-analytically in its interpretive openness and potentiality-for-being. In this sense, it is something like a "mysterious act".

Apart from Heidegger's existential conception, one might argue that the mathematical projection – when approached in terms of scientific practices – is something that has its important place in the structure of scientific inquiry, but it is by no means to be treated on the same level on which the existential phenomena are addressed with regard to the temporality of existence. In treating the mathematical projection in this way, Heidegger fails to cope with the following paradoxical situation: The mathematical determination of scientific inquiry is existentially conditioned by a certain regime of temporalizing, but the basic structure on which this determination is predicated – the mathematical projection – is without its own existential genesis which leads to mathematical essentialism within the scope of existential analytic. Heidegger postulates that this projection takes place on the level of the existential phenomena, but he attentively avoids to regard it as an existential phenomenon. He has good reasons for doing so: The mathematical projection – in contrast to the "existentiales" – is not characterized by interpretive openness and potentiality-for-being. Kochan makes a step further in admitting that the mathematical projection is an existential phenomenon. In so doing, he commits a basic mistake. Because of it he probably attributes a kind of "mathematical existentialism" to the existential conception. This mistake is also related to another shortcoming in his considerations I already discussed: Had he regarded the mathematical projection of nature not as an existential-ontological event, he would have avoided the unhappy proposal to integrate historical and sociological explanations in existential phenomenology. (The treatment of the mathematical projection is really in need of such explanations just because this projection is not part and parcel of hermeneutic-phenomenological theory of meaning constitution. By incorporating the mathematical projection in this theory, Heidegger attains the two things which I already mentioned – (a) the dispensability of a phenomenological analysis of its existential genesis; and (b) the immunization of the concept to be spelled out in historiographical, cultural-theoretical, cognitive-psychological, sociological or whatever empirically relevant terms. Again, the "special status" assigned to it in the existential conception makes the mathematical projection a mysterious act.)

The status of the mathematical projection in the existential analytic can be further clarified by formulating the following "demand of translatability" which takes into account Heidegger's ontological difference: Only what (in existential analytic) is scrutinized ontologically in its interpretive openness and potentiality-for-being might be amenable to be spelled out in ontic terms and addressed by empirical studies of disciplinary discourses. The foregoing discussion shew that Heidegger's mathematical projection meets in no way this demand. Since the demand of translatability is not met, Heidegger's mathematical projection is not to be approached empirically. As I stressed, the grounding concept of the existential conception is doomed to remain untranslatable in ontic terms. In this regard, I am wondering how Jeff Kochan as a sociologist and social historian of science would empirically identify Heidegger's mathematical projection of nature, before submitting it to whatever explanations. Is the mathematical projection of nature – as a dimension of modern epoch in the history of being – not predicated on its own cultural-historical dynamics? Is it not a lasting cultural process – persisting more than three centuries – that has to be scrutinized step-by-step (in the manner in which, for instance, Alexander Koyré has done this) in order to shed light on the cultural-historical being of modern scientific objectification of the world? By making the mathematical projection a mysterious act within the existential analytic and stylizing it as a dimension of the epochal history of being, Heidegger too easily binds this projection to metaphysics of presence.

On cognitive existentialism, science objectifies the world through the idealizing mathematical projection. Yet the thematic objectification of the world takes place within the world-as-horizon. By implication, the projection which objectifies does not destruct or rule out the interpretative openness of what becomes thematically objectified. To put it otherwise, the thematically objectified world (and its mathematical projection) are still involved in the circularity of (scientific) practices and possibilities (for doing research). The empirical identification of the mathematical projection is to be accomplished with regard to *sui generis* empirical character of this circularity-as-facticity. (Here I have in mind the distinction between the empirical as facticity [or the empirical in its situated transcendence] and the empirical as factuality. This distinction was already discussed – in the rejoinder to Gironi – in connection with the perspective on "naturalizing hermeneutic philosophy of science".)

In accordance with his treatment of the mathematical projection of nature as an existential phenomenon, Jeff Kochan ascribes to this projection a transcendental status. *Prima facie* this ascription seems intuitively justified. The mathematical projection is, as it were, a transcendental condition of formalization, theoretical conceptualization, quantification and measurements through calibrated instruments, theoretical codification of phenomena as theory's empirical basis, deductively organized explanatory scenarios, verifiable theoretical predictions, semantic interpretation of theoretical models through data models, and several other constitutive features of modern scientific research. Cognitive existentialism suggests various lines of de-transcendentalizing the mathematical projection. The latter is not an a priori condition for having scientific research as a mode of existence. The mathematical projection which discloses a region of scientific discovery by means of research practices consists itself of practices characterized by regimes of temporalizing/historizing, and accordingly, distinguished by empirically conceivable historical dynamics. This projection might only be accomplished by practices of idealizing formalization that interfere with other configured practices of scientific research. The disclosure of a domain takes place in a characteristic hermeneutic situation within an open horizon of possibilities. The mathematical projection is a facet of such a situation. Even the principles of symmetry and invariance of the fundamental scientific theories are not "mathematical essences" without existential genesis and potentiality for being, and immune to revisions in the course of projecting and appropriating possibilities – a case well documented by history of science.

It is worth mentioning that Heidegger is by no means willing to combat the established (neo-Kantian) view about the epistemology of scientific research: Science starts with idealization and mathematical codification of experience whereby a mathematically determined domain becomes opened for empirical (experimental) discoveries. The most prominent exponent of this Kantian structuralism – which in this case is tantamount to Kantian epistemological essentialism – about science in the 1920s and early 1930s was Ernst Cassirer. For him, the architectonic of symmetry principles, mathematically codified laws, and results of measurements (data-models) delimits a domain within objective reality, granted that the boundaries of the whole objective reality are delimited by the whole architectonic of science. Since a formalism that determines what is to be empirically discovered describes/represents the domain up to isomorphism, it does not specifies particular objects but refers to domain's whole structure. French and Ladyman (2003) could have been equally successful if they have been latched their ontic structural realism not onto Cassirer's structuralism, but onto Heidegger's concept of science's objectifying thematization by means of mathematical projection. When Heidegger on various occasions argues that contemporary (relativistic and quantum) physics – though re-describing nature entirely in new terms – does not invalidate classical mechanics, he actually supports the central claim of scientific structuralism – cumulative development of science with regard not to empirical knowledge but mathematical structures. What retained after a scientific revolution related to a global theory-change is structure.)

To sum up, in contrast to the existential conception cognitive existentialism claims that science does not have epistemological, mathematical, methodological, transcendental or whatever essence. Science produces interpretively (in the process of inquiry) its own cognitive specificity within its potentiality-for-being. Being manifested by a variety of hermeneutic circles, this specificity never becomes a pure presence ready to be schematized/fixed in epistemological, semantic, or cognitive-sociological terms.

E

I did my best to understand what exactly Kochan's notion of "mathematical existentialism" should mean. Unfortunately, my efforts were in vain. Kochan is highly parsimonious in instructing us how to understand the mathematical projection of nature within the horizon of mathematical existence.

Historically seen, there is a highly interesting hermeneutic-phenomenological theory of mathematical existence. Its author is Husserl's pupil Oskar Becker who radicalizes Heidegger's existential ontology in several respects in order to make it pertinent to the constitution of (transfinite) mathematical objects. In fact, Becker's theory is not an isolated and exotic development of phenomenology. Its creation took place within a growing dialogue between mathematical constructivism and phenomenological philosophy. In the 1920s the intuitionists made serious efforts to show that all mathematical objects are (mental) constructions achieved by human beings not in an atemporal manner but with respect to the temporal duration of the constructive acts. On this account, the finite existence of a mathematician is not capable to construct most of the infinite totalities whose existence is postulated by classical mathematics. Since the existence of mathematical objects is a mentally constructive process, there are no underlying logical foundations of what becomes constructed, and the constructions are not to be completely captured in formal systems. (In construing constructability phenomenologically, Arend Heyting treated these constructions as fulfillable mathematical intentions.) The idea of "mathematical existence" is inherent in what intuitionists (after Luitzen Brouwer) called "choice sequences". Roughly, the latter are introduced by doing justice to phenomenological analysis – though not in a strictly Husserlian fashion – of the intuitive continuum. (van Atten, 2005) The potential infinity of the number of choices is the principal feature of this thought tool for modelling the continuum of mathematical entities of a given sort (e.g. the continuum of rational numbers). By specifying a condition on several sequences, one is able to bundle choice sequences in a "spread", thus figuring out the mathematical entities' constructive-temporal mode of existence.

The whole approach to the intuitive continuum is designed to unfold a strong alternative to the classical programs in which the continuum is presented as an infinite set of discrete elements. At this point, one ought to indicate another significant similarity to phenomenology – the continuity of choice sequences bears a resemblance to the continuity of the temporalizing constitution of meaning in the intentional life of consciousness. The specifying condition cannot finalize a sequence of choices. Each sequence remains open with regard to future free (undetermined by axiomatic structure) choices. (It is another question that the choices might be prescribed by algorithm posed on the sequence by the mathematician. It should be mentioned that the mathematicians are very often using in their constructive work "lawless choice sequences".) To sum up, the temporalized existence of mathematical objects hinges on the mathematicians' choices and decisions.

Intuitionist temporality of the mathematical objects has an important existential-phenomenological dimension. The futural openness of these objects' existence corresponds to the temporality of construction. The concept of choice sequence alludes to a unity of the existence of the mathematical objects and the mathematicians' temporal mode of existence in which these objects are constituted. It is this unity qua "mathematical existence" that Becker spells out in terms of both Husserl's transcendental phenomenology and Heidegger's hermeneutic phenomenology, revising both of them significantly. (Ginev and Andreev, 2008; Ginev, 2009) In line with Husserl's phenomenology – and by equating propositions with intentions – he treated intuitionist constructability as a kind of constitution. Much more important than Husserl's constitutional analysis is the impact of hermeneutic phenomenology on Becker's existentialism about mathematics. This impact is manifested most typically by the insistence on the interpretive-ontological openness of all mathematical entities: "Mathematical existence" takes place in open horizons of constructability. The horizonality of this existence is not to be ruled out when one looks for access to a "mathematical state of affairs" (*mathematischer Sachverhalt*). By implication, the access – as elicited by a constructive proof – that takes into account the open horizon of constructability belongs to the respective mathematical state of affair.

Becker is particularly interested in the existence of transfinite ordinals and, more generally, entities defined by means of transfinite recursion. Heidegger's hermeneutics of facticity is about Dasein's modes of being-in-the-world as distinguished by finitude. It seems counterintuitive to admit that the finite existence that is able to experience its finitude could be liable to constitute (through constructive methods like choice sequences) transfinite entities. Nonetheless, Becker (1959, p. 154) argues that only a finite creature is able to make sense of the task of "mastering the infinite" (*Beherrschung des*

Unendlichen). In this regard, he introduces a special mode of existence that is not characterized by meta-temporality (eternity), but by a constant transcending of human finitude in constituting transfinite objects. Thus, the latter are not eternal objects existing beyond the temporalizing of temporality and freed from open horizons. The possibility of mastering the infinite in mathematics is rooted in the capability "immanent transcendence" of mathematizing as a mode of existence.

I can only speculate about whether Kochan has in mind Becker's theory when speaking of mathematical existence. Even if Kochan implicitly refers to Becker's theory, the question arises as to whether Heidegger's concept of the mathematical projection is congruent with Becker's version of mathematical existentialism. And the only possible answer sounds "not at all". I already mentioned Heidegger's tacit commitment to Husserl's philosophical extension of the mathematical notion of manifold. This extension is entirely in line with the essentialist assumptions of classical mathematics. In his earlier work Husserl tended to the view that the truth (the semantic completeness) of an axiomatized mathematical theory (considered as a "definite manifold") follows from its consistency. Though he later never discussed this (Hilbertian) view, Husserl did not give up his belief in it. Accordingly, what is only capable to exist mathematically is the entities defined/determined by the theory's axioms. Only the axiomatically determined existence of formal entities is to be allowed in mathematics. This is the view of mathematical existence Heidegger inherits of Husserl – the view which is integrated in the concept of the mathematical projection of nature. One might conclude that like Husserl, Heidegger shows respect for classical mathematics. A radical – intuitionist or non-intuitionist – constructivism (that in a meta-mathematical context would gravitate to a kind of "existentialism") threatens the whole architectonics of classical mathematics. Yet it is hardly to believe that such considerations have been crucial for Heidegger. (Even in his "longest day" he has never been a philosopher of mathematics.) His aversion to Becker-like mathematical existence is evident: "Mathematical existentialism" simply contradicts the scenario of the existential conception of science. This existentialism would transform the mathematical projection from a determining essence (determination of existence by a formal structure) into an open horizon of potentially infinite constructability.

Finally, let me point out that I am not trying to suggest that the only path to "mathematical existentialism" is via Becker's extension of hermeneutic phenomenology. To be sure, one can make sense of the concept of mathematical existence through a pertinent reading of Wittgenstein's philosophy of mathematics as well. But there is no indication in Kochan's paper that he is willing to follow the lead of this philosophy. And what is more important, in the framework of Wittgenstein's philosophy of mathematics there is no room for "the mathematical projection of nature" even if one goes on to assume that this projection elicits a new "form of life".

Heidegger's existential conception of science is the birthplace of all hermeneutic studies of scientific inquiry. Criticizing this conception is a kind of thinking with Heidegger against Heidegger. The whole program of cognitive existentialism is predicated on this kind of thinking.

NOTES

3. See on this argument Vogel (1991).
4. This talk of "external things" indicates how strongly Lonergan is committed to the primacy of the epistemic cut.
5. On the reception of this discovery see Kohler (1972). The first oxidative enzyme had already been found (by Yoshida) before Buchner's discovery.
6. On the profound impact of Buchner's discovery on the development of enzymology see Bechtel and Abrahamsen (2007, 6).
7. See in this regard Kohler (1972, 327).
8. See Kohler (1973); Fruton (1999, 144–160).
9. See Holmes (1992, 54-58).
10. This status is a condition for the possibility of making theoretical predictions that serve the function of a theory's independent tests.
11. Here I am using the expression "spaces of representation" in Rheinberger's (1995) sense.
12. In what follows I draw on Allchin's (1997) case study.
13. In fact, in *Being and Time* Heidegger does not give an account of why science, whose "theoretical attitude is without circumspection", should present an authentic mode of existence. See Ginev (2011, 10–11).
14. For various reasons, I prefer to translate Heidegger's *umsichtige Überlegung* as "deliberative circumspection" and not as "circumspective deliberation".
15. I follow Rouse's (2002, ch. 8) analysis at this point.
16. For these scenarios see Bokulich (2011).
17. See Polikarov (1974), Ginev (1986), Ginev (1997a), Ginev (2003).
18. On my proposal for how to make Heidegger's Dasein relevant to the social theory based on a trans-subjective view of practices see Ginev (2014a).
19. In most of the cases in which Kochan speaks of existence he means possibly facticity. It seems to me that he does not take into account Heidegger's threefold distinction between existence, facticity, and fallenness.
20. See Zilsel (2003), and for a critical analysis Krohn (1985).
21. See Heidegger (1962, 410).
22. See in particular Heidegger (1984, 98), and Heidegger (1993, 296).

WORKS CITED

Agamben, Giorgio. (1998) *Homo Sacer. Sovereign Power and Bare Life*. Stanford: Stanford University Press.

Agre, Philip E. (1997) *Computation and Human Experience*. Cambridge, England: Cambridge University Press.

Allchin, Douglas. (1997) "A Twentieth Century Phlogiston," *Perspectives on Science* 5, pp. 81–127.

Babich, Babette. (2002) "The Fortunes of Incommensurability: Thoughtstyles, Paradigms, and Patrck A. Heelan's Hermeneutics of Science." In *Hermeneutic Philosophy of Science, Van Gogh's Eyes, and God*, ed. Babette Babich. Dordrecht: Kluwer, pp. 1–18.

————. (2010) "Towards a Critical Philosophy of Science: Continental Beginnings and Bugbears," *International Studies in the Philosophy of Science* 24:4, pp. 343–391.

———. (2014) "The Multidimensionality of Hermeneutic Phenomenology: From Philology through Science and Technology to Theology." In *The Multidimensionality of Hermeneutic Phenomenology*, eds. Babette Babich and Dimitri Ginev. Dordrecht: Springer, pp. xv–xxxv.

Bechtel, William. (1984) "Reconceptualizations and interfield connections: The discovery of the link between vitamins and coenzymes," *Philosophy of Science* 51, pp. 265–292.

Bechtel William and Abrahamsen, Adele. (2007) "In Search of Mitochondrial Mechanism:Interfield Excursions between Cell Biology and Biochemistry," *Journal of the History of Biology* 40, pp. 1–33.

Becker, Oskar. (1959) *Größe und Grenze der mathematischen Denkweise*. Freiburg und München: Alber.

Bloor, David. (1992) "Left and Right Wittgensteinians." In *Science as Practice and Culture*, ed. Andrew Pickering. Chicago and London: The University of Chicago Press.

———. (1999) "Anti-Latour," *Studies in History and Philosophy of Science* 30:1, pp. 81–112.

Brading, Katherine and Elena Castellani. (2003) "Introduction." In *Symmetries in Physics*, eds. Katherine Brading and Elena Castellani. Cambridge: Cambridge University Press, pp. 1–18.

Brandom, Robert. (1997) "Study Guide." In Wilfrid Sellars, *Empiricism and the Philosophy of Mind*. Cambridge, Massachusetts and London, England: Harvard University Press, pp. 119–181.

Bokulich, Alisa and Bokulich, Peter, eds. (2011) *Scientific Structuralism*. Dordrecht: Springer.

Brooks, Rodney A. (1999) *Cambrian Intelligence. The Early History of the New AI*. Cambridge, Massachusetts and London, England: The MIT Press.

Centrone, Stephania. (2009) *Logic and Philosophy of Mathematics in the Early Husserl*. Dordrecht: Springer.

Chang, Hasok. (2011) "The Philosophical Grammar of Scientific Practice," *International Studies in the Philosophy of Science* 25:3, pp. 201–221.

Cox, Christoph. (1999) *Nietzsche. Naturalism and Interpretation*. Berkeley, Los Angelis, Oxford: University of California Press.

Crease, R. (2009) "Covariant Realism", *Human Affairs* 19, pp. 223–232.

———. (2012) "Heidegger and the Empirical Turn in Continental Philosophy of Science." In *Heidegger on Science*, ed. Trish Glazebrook. Albany: State University of New York Press, pp. 225–238.

Darden, Lindley and Maull, Nancy. (1977). "Interfield Theories," *Philosophy of Science* 44:1, pp. 43–64.

Dreyfus, Hubert (2007) "Why Heideggerian AI Failed and How Fixing It Would Require Making It More Heideggerian," *Philosophical Psychology* 20:2, pp. 247–268.

Esfeld, Michael and Lam, Vincent. (2011) "Ontic Structural Realism as a Metaphysics of Objects." In *Scientific Structuralism*, eds. Alisa Bokulich and Peter Bokulich. Dordrecht: Springer, pp. 143–160.

French, Steven and Ladyman, James. (2003) "Remodelling structural realism: Quantum physics and the metaphysics of structure," *Synthese* 136, pp. 31–56.

Fruton, Joseph S. (1999) *Proteins, Enzymes, Genes: The Interplay of Chemistry and Biology*. New Haven and London: Yale University Press.

Fruton, Joseph S. (2006) *Fermentation: Vital or Chemical Process?* Leiden: Brill.

Gadamer, Hans-Georg. (1989) *Truth and Method*. Second edition with translation revisited by J. Weinsheimer and D.G. Marshall. London and New York: Continuum.

Gavroglu, Kostas. (1995) *Fritz London – A Scientific Biography*. Cambridge: Cambridge University Press.

Ginev, Dimitri. (1986) "The 'Science-Methodology' Iterative Circle," *Journal for General Philosophy of Science* 17:1, pp. 143–153.

————. (1997a) "On the Dialectical Nature of Polikarov's Philosophical Methodology of Science." In *Issues and Images in the Philosophy of Science. Scientific and Philosophical Essays for Azarya Polikarov*, eds. Dimitri Ginev and Robert S. Cohen. Dordrecht, Boston, London: Kluwer, pp. xiii–xviii.

————. (1997b) *A Passage to the Hermeneutic Philosophy of Science*, Amsterdam and Atlanta, GA: Rodopi.

————. (2000) *Critique of Epistemological Reason*. Sofia: Bulgarian Academic Monographs.

————. (2003) "Azarya Polikarov's 'Heuristic Realism'," *Epistemologia* 25:2, pp. 311–318.

————. (2005) Ägainst the Politics of Postmodern Philosophy of Science." *International Studies in the Philosophy of Science* 19:2, pp. 191–208.

————. (2009) "Interpretative Erschlossenheit der endlichen Existenz und mathematische Unendlichkeit," *Studia Phaenomenologica* 9: pp. 495–508.

————. (2011) *The Tenets of Cognitive Existentialism*. Athens, OH: Ohio University Press.

————. (2012) "Two Accounts of the Heremenutic Fore-structure of Scientific Research," *International Studies in the Philosophy of Science* 26:4, pp. 423–445.

————. (2013a) "Scrutinizing Scientism from a Hermeneutic Point of View," *Social Epistemology* 27:1, pp. 68–89.

————. (2013b) "Scientific Structuralism in the Time of Poststructuralism," *Divinatio. Studia Culturologica* 37:1, pp. 221–238.

————. (2013c) "Hermeneutic Pre-Normativity," *Philosophical Forum*, 44:3, pp. 275–293.

————. (2014a) "Social practices from the viewpoint of trans-subjective existentialism," *European Journal of Social Theory* 17:1, pp. 77–94.

————. (2014b) "Textualizing beyond Rorty's textualism," *South African Journal of Philosophy* 33:3, pp. 285–296.

Ginev, Dimitri and Andreev, Jassen. (2008) "Die mathematische Existenz aus phänomenologischer Sicht." In *Aspekte der phänomenologischen Theorie der Wissenschaft*, Hg. Dimitri Ginev. Würzburg: Königshausen & Neumann, pp. 136–154.

Haugeland, John. (2013) *Dasein Disclosed*. Edited by John Rouse. Cambridge, Massachusetts and London, England: Harvard University Press.

Heidegger, Martin. (1962) *Being and Time*. Translated by John Macquarrie & Edward Robinson. San Francisco: Harper Collins Publishers.

————. (1977) *The Question Concerning Technology and Other Essays*. Translated by William Lovitt. New York and London: Garland Punlishing.

————. (1984) *Die Frage nach dem Ding (Zu Kants Lehre von den transzendentalen Grundsätzen)*. Freiburger Vorlesung Wintersemester 1935/36, Hg. Petra Jager, GA 41. Tübingen: Max Niemeyer.

————. (1994) *Basic Writings*. Edited by David Farrell Krell. London: Routledge.Holmes, Frederic C. (1992) *Between Biology and Medicine: The Formation of Intermediary Metabolism*. Berkeley, CA: Office for History of Science and Technology, University of California at Berkeley.

Holmes, Frederic (1992), *Hans Krebs - The Formation of Scientific Life (1900–1933)*, Vol.1. Oxford and New York: Oxford University Press.

Kochan, Jeff. (2012) "Review of *The Tenets of Cognitive Existentialism*," *Notre Dame Philosophical Reviews* 2012.04.23.

———. (2014) "Freedom, Forgetting and Solidarity." In *Forgetting*, eds. Giovani Galizia and David Shulman. Jerusalem: Magnes Press, pp. 216–218.

Kockelmans, Joseph J. (1985) *Heidegger and Science*. Washington, D.C.: Center for Advanced Studies in Phenomenology and University Press of America.

Kohler, Robert. (1972) "The Reception of Eduard Buchner's Discovery of Cell-Free Fermentation," *Journal of the History of Biology* 5:2, pp. 327–353.

———. (1973) "The enzyme theory and the origins of biochemistry," *Isis* 64, pp. 181–196.

Kornblith, Hilary. (2007) "The Naturalistic Project in Epistemology: Where Do We Go from Here?" In In *Naturalized Epistemology and Philosophy of Science*, eds. Chienkuo Michael Mi and Ruye-lin Chen. Amsterdam and New York: Rodopi, pp.39–60.

Kosso, Peter. (2003) "Symmetry, objectivity, and design." In *Symmetries in Physics*, eds. Katherine Brading and Elena Castellani. Cambridge: Cambridge University Press, pp. 412–424.

Krohn, Wolfgang. (1985) "Edgar Zilsel zur Methodologie einer exakten Geisteswissenschaft." In *Philosophie, Wissenschaft, Aufklärung. Beiträge zue Geschichte und Wirkung des Wiener Kreises*, Hg. H.-J. Dahms. Berlin & New York: W. de Gruyter, pp. 257–275.

London, F. (1923). "Über die Bedingungen der Möglichkeit einer deduktiven Theorie", *Jahrbuch für Philosophie und phänomenologische Forschung*, VI, 335-384.

Lonergan, Bernard. (1992) *Insight: A Study of Human Understanding*. Edited by F.E. Crowe and R.M. Doran, Toronto/Buffalo/London: University of Toronto Press.

Pickering, Andrew. (1995) *The Mangle of Practice: Time, Angency and Science*. Chicago: Chicago University Press.

Polikarov, Azarya. (1974) "The Divergent-Convergent Method – a Heuristic Approach to Problem-Solving." In *Methodological and Historical Essays in the Natural and Social Sciences*, eds. Robert S. Cohen and Marx W. Wartofsky. Dordrecht and Boston: Reidel,pp. 211–233.

Rheinberger, Hans-Jörg. (1995) "From Microsomes to Ribosomes: 'Strategies' of 'Representation'," *Journal of the History of Biology* 28:1, pp. 49–89.

Rorty, Richard. (1979) *Philosophy and the Mirror of Nature*. Princeton: Princeton University Press.

Rouse, Joseph. (2002) *How Scientific Practices Matter*. Chicago: University of Chicago Press.

———. (2007) "Naturalism and Scientific Practices: A Concluding Scientific Postscript." In *Naturalized Epistemology and Philosophy of Science*, eds. Chienkuo Michael Mi and Ruye-lin Chen. Amsterdam and New York: Rodopi, pp. 61–86,

Theodorou, Panos (2012). "Cognitive Existentialism, Phenomenology, and Philosophy of Science: Stimulating the Dialogue," *International Studies in the Philosophy of Science*, 26:3, pp. 335–343.

van Atten, Mark. (2005) "Phenomenology's Reception of Brouwer's Choice Sequences." In *Oskar Becker und die Philosophie der Mathematik*, Hg. Annemarie Gethmann-Siefert.München: Wilhelm Fink, pp. 101–118.

Vogel, Steven. (1991) "Science, practice and politics," *Social Epistemology* 5, pp. 267–292.

Winograd, Terry and Flores, Fernando. (1986) *Understanding Computers and Cognition*. Norwood: Ablex Publishing Corporation.

Zahnen, Barbara. (2011) "Vollzug und Sprache Physischer Geographie und die Frage geographischen Takts," *Social Geography*, 6:1, pp. 47–61.

Zilsel, Edgar. (2003) *The Social Origins of Modern Science*. Edited by Diederick Raven, Wofgang Krohn. Dordrecht: Springer.

ABOUT THE AUTHORS

ROBERT CREASE is Professor of Philosophy, Chairman of the Department of Philosophy at Stony Brook University. He is a Fellow of the American Physical Society, and the Institute of Physics in London. Among his books are: *World in the Balance: The Historic Quest for an Absolute System of Measurement* (2011), *The Philosophy of Expertise* (with E. Selinger 2006), *Making Physics: A Biography of Brookhaven National Laboratory 1947–1972* (1999), and *The Play of Nature: Experimentation as Performance* (1993). He edited *Hermeneutics and the Natural Sciences* (1997).

DIMITRI GINEV is a political philosopher, and the founder of the journal “Studia Culturologica. Divinatio”. He has taught at several European and American universities. His books include *Models of Science's Development* (1986), *Dialogues on Scientific Rationality* (1989), *Grundriss einer kritischen Wissenschaftstheorie* (1989), *Essays in Hermeneutics of Science* (1997), *Entre Anthropologie et Hermeneutique* (2003), *Steps in Overcoming Scientism* (2007), *Transformationen der Hermeneutik* (2008). With Robert Cohen he edited *Issues and Images in the Philosophy of Science* (1997), and with Babette Babich *The Multidimensionality of Hermeneutic Phenomenology* (2014).

FABIO GIRONI is an IRC Postdoctoral Research Fellow in the School of Philosophy, University College Dublin. He previously studied at the University of Rome “La Sapienza”, the University of London, and Cardiff University, where he obtained his Ph.D. His work focuses on the philosophy of science and history of scientific conceptual frameworks, drawing from both analytic and Continental sources. He is the editor of the journal *Speculations*.

EVALDAS JUOZELIS is Doctor of Philosophy, and Assistant Professor at the Institute of Philosophy and Humanities at the Mikolas Romeris University in Vilnius, Lithuania. He has published widely on issues related to the theory of practices, and the philosophy of science based on this theory.

JEFF KOCHAN is currently Substitute Full Professor of Science and Technology Studies at the University of Freiburg. He is also an Associated Research Fellow in the Zukunftskolleg at the University of Konstanz. His work combines existential phenomenology with topics in the history, philosophy, and sociology of science and technology. He is near completion a book manuscript titled *Science as Social Existence: An Essay on Heidegger and the Sociology of Scientific Knowledge*. For a publication list, see: http://philpapers.org/s/Jeff.

EMIL LENSKY is Doctor of Human Ecology and member of the research group Documenta. He focuses on the dialogue between the hermeneutic philosophy of science and holistic approaches in theoretical biology. Drawing on ideas developed in deep ecology and green anarchism, he became an activist of several organizations devoted to the animal rights. A study of him on the hermeneutic roots of cognitive existentialism can be found in *Philosophische Rundschau*, 2012.

ARVIN VOS is Professor of Philosophy at Western Kentucky University, Bowling Green. His specialty is the philosophical culture of the Middle Ages, with special interest in Augustine and Aquinas. He is the author of *Aquinas, Calvin, & Contemporary Protestant Thought. A Critique of Protestant Views on the Thought of Thomas Aquinas* (1985), among many other works. In drawing on Bernard Lonergan's work, he has unfolded a heuristic epistemology of scientific research.

INDEX